Inclusive Designing

P. M. Langdon · J. Lazar · A. Heylighen
H. Dong
Editors

Inclusive Designing

Joining Usability, Accessibility, and Inclusion

 Springer

Editors
P. M. Langdon
Department of Engineering
   Cambridge Engineering
   Design Centre
University of Cambridge
Cambridge
UK

J. Lazar
Universal Usability Laboratory
Department of Computer
   and Information Sciences
Towson University
Towson, ML
USA

A. Heylighen
Department of Architecture
   Urbanism and Planning
Katholieke Universiteit Leuven
Heverlee, Leuven
Belgium

H. Dong
College of Design and Innovation
Tongji University
Shanghai
China

ISBN 978-3-319-05094-2   ISBN 978-3-319-05095-9   (eBook)
DOI 10.1007/978-3-319-05095-9
Springer Cham Heidelberg New York Dordrecht London

Library of Congress Control Number: 2014933669

Printed on acid-free paper

Springer is part of Springer Science+Business Media (www.springer.com)

Preface

The Cambridge Workshops on Universal Access and Assistive Technology (CWUAAT) are a series of workshops held at a Cambridge University College every two years. This volume: "Inclusive Designing: Joining Usability, Accessibility, and Inclusion" comes from the 7th in this series of highly successful events, held March 2014 at the University of Cambridge.

The workshops are characterised by a single session running over 3 days in pleasant surroundings with delegates from home and abroad staying on site. Allowing speakers longer presentation times, carrying discussion on through sessions into plenaries and shared social and leisure spaces generates an enjoyable academic environment that is both creative and innovative. CWUAAT is one of the few gatherings where people interested in inclusive design, across different fields, including *designers, computer scientists, engineers, architects, ergonomists, policymakers and user communities, meet, discuss and collaborate*. CWUAAT has also become a much more international workshop, representing diverse cultures including India, China, Slovakia, USA, Belgium, UK, Denmark and many more. CWUAAT has worked towards the goal of being internationally inclusive. In addition, the doctoral consortium has successfully returned to the programme in 2014.

Inclusive Design Research involves developing tools and guidance enabling designers to design for the widest possible population, for a given range of capabilities. In the context of emerging demographic changes leading to greater numbers of older people and people with disabilities, the general field of inclusive design research strives to relate the design of products to the capabilities of the population. Inclusive populations of older people contain a greater variation in sensory, cognitive and physical user capabilities. These variations may be cooccurring and rapidly changing which leads to a demanding design environment. Inclusive Design Research involves developing methods, technologies, tools and guidance for supporting product designers, software developers and architects to design for the widest possible population for a given range of capabilities, within the contemporary social and economic context.

Recent research developments have addressed these issues in the context of wireless communications, governance and policy, daily living activities, the workplace, the built environment, interactive digital TV, computer gaming and mobile devices. This now reflects the multi-disciplinary approach that is required for the diverse, sometimes conflicting demands of design for ageing and impairment, usability and accessibility and universal access. CWUAAT provides a platform for such a need. The workshop focused on six main themes:

I *Measuring Product Demand and Peoples' Capabilities*
Measuring users' capabilities has always been central to Inclusive Design. Biswas and Langdon extend this into a large developing world context by carrying out and analysing a survey of Indian users in a comparative study. Choi highlights the weaknesses of conventional ergonomic measures for predicting manual dexterity.

II *Designing Cognitive Interaction with Emerging Technologies*
The workshop now covers a wide range of cognitive science, from testing the use of advanced user modelling as an alternative to user trials, to manipulating the motivation of software users on a gender basis. This is without losing sight of its assistive technology roots, as papers here still represent: cursor manipulation, adaptive interfaces and the inclusive advantages of alternative operating system interfaces.

III *Reconciling Usability, Accessibility and Inclusive Design*
It is not enough to aim for improved usability or accessibility by simply following a set of design guidelines, without listening directly to potential users. Inclusive designing means both using existing knowledge and design guidelines and involving stakeholders throughout the design process, to improve the usability and accessibility of interfaces and products. Horton and Sloan document techniques for improved lifecycle models that incorporate policies, guidelines and inclusion. Chakraborty, Hritz and Dehlinger investigate design requirements for computer gaming by blind users, by collecting survey and usability testing feedback from six visually impaired gamers. Heitor et al. conducted a case study at a university campus, with people with various perceptual and motor impairments and analysed the physical accessibility of the campus, revealing major problems.

IV *Designing Inclusive Spaces*
Contributions to CWUAAT for the architectural field are now so established that they manifest an eclectic range of methods and topics, within the broad constraint of buildings and spaces. For example, the journey to the operating room; surveying an inclusive educational environment in Bratislava; sampling the views of residents of 35 care homes in London; and charting the differences of geographically distributed public transport train companies, all come within its scope.

V *Collaborative and Participatory Design*
A common theme heard from disability advocates is 'nothing for us, without us'. At the core, inclusive design is not only about designing for diversity, but also designing with diverse groups, ensuring that the user populations themselves drive the design process. Researchers and practitioners continue to look for new approaches and methods to more successfully manage the

design process while including diverse voices. All the papers in this section illustrate both breadth and depth of this approach.

VI *Legislation, Standards and Policy*

Researchers are increasingly investigating how public policies (both from governments and international non-governmental organisations) influence inclusive and accessible design. Wentz and Lazar describe how, due to changes in how government provides services, citizens are often coming to public libraries for help, and their chapter describes the inclusive design process for online resources for public librarians. Gomez et al. present considerations behind comparative studies of barriers to employment in Chile and the UK, raising intriguing issues of the existence of religious models of disability.

This book contains the best reviewed papers from CWUAAT 2014 that were invited for oral presentation. The papers that have been included were selected by blind peer review carried out by an international panel of currently active researchers. The chapters forming the book represent an edited sample of current national and international research in the fields of inclusive design, universal access and assistive and rehabilitative technology.

We would like to thank all those authors and contributors who have contributed to CWUAAT 2014 and to the preparation of this book. We would also like to thank the additional external reviewers who took part in the review process. Many thanks are also due to the reviewing members of the Programme Committee who continue to support the workshop series. Finally, Mari Huhtala and Anthea Maybury do a superb professional job in taking the final submissions through to publication-ready manuscript in time for the Workshop itself. We are grateful to the staff at Fitzwilliam College for their patience and help.

March 2014
                                                                              Pat Langdon
                                                                              Jonathan Lazar
                                                                              Ann Heylighen
                                                                              Hua Dong

Contents

Part III   Reconciling Usability, Accessibility and Inclusive Design

Part IV   Designing Inclusive Spaces

Contributors

C. Andrews The National Centre for Product Design and Development Research, PDR, Cardiff Metropolitan University, Cardiff, UK

M. Annemans Osar Architects nv, Department of Architecture, Urbanism and Planning, Katholieke Universiteit Leuven, Leuven, Belgium

S. Baumers Department of Architecture, Urbanism and Planning, Katholieke Universiteit Leuven, Leuven, Belgium

A. E. Beckett School of Sociology and Social Policy, Centre for Disability Studies, University of Leeds, Leeds, UK

J. A. Bichard Department of Engineering, Engineering Design Centre, University of Cambridge, Cambridge, UK

P. Biswas Department of Engineering, Engineering Design Centre, University of Cambridge, Cambridge, UK

Z. Ceresnova Faculty of Architecture, Centre of Design for All, Slovak University of Technology in Bratislava, Bratislava, Slovakia

J. Chakraborty Universal Usability Laboratory, Department of Computer and Information Sciences, Towson University, Towson, USA

Y. M. Choi School of Industrial Design, Georgia Institute of Technology, College of Architecture, Atlanta, USA

P. J. Clarkson Department of Engineering, Engineering Design Centre, University of Cambridge, Cambridge, UK

S. Cook User-Centred Design Group, Loughborough Design School, Loughborough University, Loughborough, UK

J. Dehlinger Universal Usability Laboratory, Department of Computer and Information Sciences, Towson University, Towson, USA

H. Dong Inclusive Design Research Centre, College of Design and Innovation, Tongji University, Shanghai, China

L. V. L. Filgueiras Polytechnic School of Engineering, University of São Paulo, São Paulo, Brazil

T. S. Goldhaber Department of Engineering, Engineering Design Centre, University of Cambridge, Cambridge, UK

J. P. Gomez Department of Engineering, Engineering Design Centre, University of Cambridge, Cambridge, UK

T. Heitor Department of Civil Engineering, Architecture and Georesources, School of Engineering, University of Lisbon, Lisbon, Portugal

R. Herriott Design Platform, Aarhus School of Architecture, Aarhus, Denmark

A. Heylighen Department of Architecture, University of Leuven (KU Leuven), Leuven, Belgium

R. J. Holt School of Mechanical Engineering, University of Leeds, Leeds, UK

S. Horton Accessible User Experience and Design, The Paciello Group, Nashua, USA

J. Hritz Universal Usability Lab, Department of Computer and Information Sciences, Towson University, Towson, USA

E. Karanastasi Department of Architecture, Katholieke Universiteit Leuven, Leuven, Belgium

M. Kinnaer Department of Architecture, Urbanism and Planning, Katholieke Universiteit Leuven, Leuven, Belgium

S. Kittusami IITM's Rural Technology and Business Incubator, Chennai, India

P. M. Langdon Department of Engineering, Engineering Design Centre, University of Cambridge, Cambridge, UK

J. Lazar Radcliffe Institute for Advanced Study, Harvard University, Cambridge, USA; Department of Computer and Information Sciences, Towson University, Towson, USA

L. Luque Department of System Analysis and Development, São Paulo State Technological College (FATEC), Mogi das Cruzes, Brazil

V. Medeiros Faculty of Architecture and Urbanism, University of Brasilia, Brasilia, Brazil

A.-M. Moore School of Mechanical Engineering, Institute of Design, Robotics and Optimisation, University of Leeds, Leeds, UK

J. Morris Rehabilitation Engineering Research Center for Wireless Technologies, Atlanta, USA

J. Mueller Rehabilitation Engineering Research Center for Wireless Technologies, Atlanta, US

J. Murin Faculty of Electronical Engineering, Department of Computer Graphics and Interaction, Czech Technical University in Prague, Prague, Czech Republic

R. Nascimento Architecture Unit, School of Engineering, University of Lisbon, Lisbon, Portugal

G. C. Pereira Department of System Analysis and Development, São Paulo State Technological College, Mogi das Cruzes, Brazil

O. Polacek Faculty of Electronical Engineering, Department of Computer Graphics and Interaction, Czech Technical University in Prague, Prague, Czech Republic

S. Prashant IITM's Rural Technology and Business Incubator, Chennai, India

G. Raheja Department of Architecture and Planning, Indian Institute of Technology Roorkee, Roorkee, India

S. A. Shamshirsaz Inclusive Design Research Group, School of Engineering and Design, Brunel University, London, UK

D. Sloan The Paciello Group, London, UK

A. J. Sporka Faculty of Electronical Engineering, Department of Computer Graphics and Interaction, Czech Technical University in Prague, Prague, Czech Republic

S. Suryawanshi Department of Architecture and Planning, Indian Institute of Technology Roorkee, Roorkee, India

A. Tomé Department of Civil Engineering, Architecture and Georesources, Instituto de Engenharia de Estruturas, Território e Construção, Instituto Superior Tecnico, University of Lisbon, Lisbon, Portugal

J. Umadikar IITM's Rural Technology and Business Incubator, Chennai, India

C. Van Audenhove LUCAS Center for Care Research and Consultancy, and Academic Center for General Practice, Katholieke Universiteit Leuven, Leuven, Belgium

E. S. Veriscimo Department of System Analysis and Development, São Paulo State Technological College, Mogi das Cruzes, Brazil

H. Vermolen Osar Architects nv, Antwerp, Belgium

B. Wentz Department of Management Information Systems, Shippensburg University, Shippensburg, USA

P. K. A. Wollner Department of Engineering, Engineering Design Centre, University of Cambridge, Cambridge, UK

S. Yuan Tongji University, Shanghai, China

Part I
Measuring Product Demand and Peoples' Capabilities

An HCI Survey on Elderly Users in India

P. Biswas and P. M. Langdon

1 Introduction

Human–computer interaction (HCI) is about knowing the user, which becomes more important when we consider users with different capabilities in a developing country. This paper reports a survey to estimate Indian elderly and disabled users' perceptual, cognitive and motor capabilities and also their experience and attitude towards technology. We initially identified functional parameters (Biswas and Langdon 2012; Biswas and Robinson 2013) that can affect users' interaction with electronic devices and combined both objective metrics on functional parameters and subjective attitudes towards technology. Previous surveys either concentrated on ergonomics or demographic details of users in European countries (Langdon and Thimbleby 2010) or focused on a particular device like digital TV or mobile phones (ITU 2011; Narasimhan and Leblois 2012). There is not much reported work on the capabilities and attitude towards technology of the older Indian population, especially from a HCI point of view.

It is well known that there are some social, economical and cultural differences between people living in different countries in Europe (Mackenbach et al. 1997; Siegrist and Marmot 2004; Dalstra et al. 2005). These differences have a direct influence on people's life course and their circumstances in old age. So we used data from a previous survey (Guide 2011) to compare the results of the Indian population with their European counterparts.

Our study found that there is a significant effect of age on the hand strength of elderly users limiting their use of standard computer peripherals. It is also found that European elderly users tend to score higher in cognitive tests than their Indian counterparts, and in the Indian population there is a significant correlation between education level and cognitive abilities. We also found that elderly people

P. Biswas (✉) · P. M. Langdon
Engineering Design Centre, Department of Engineering, University of Cambridge, Cambridge, UK
e-mail: pb400@eng.cam.ac.uk

P. M. Langdon et al. (eds.), *Inclusive Designing*, DOI: 10.1007/978-3-319-05095-9_1,
© Springer International Publishing Switzerland 2014

3

acknowledge the need to use new technologies though they prefer to use TV and mobile phones than computers. The paper also points out the implication of the findings for designers in Sect. 4.

2 Survey

We conducted a survey to estimate users' perceptual, cognitive and motor capabilities and also their experience and attitude towards technology. Previous surveys either concentrated on ergonomics or the demographic details of users or focused on a particular device like digital TV or mobile phones. We initially identified the users' functional parameters (Biswas and Langdon 2012; Biswas et al. 2012; Biswas and Robinson 2013) that can affect their interaction with electronic devices and combined both objective metrics on functional parameters and subjective attitudes towards technology.

2.1 Place of Survey

The survey was conducted at Mandi, Himachal Pradesh, Kolkata, West Bengal and Bangalore, Karnataka. The survey was conducted at old age homes with volunteer participants. We also collected data from nine young people with physical impairment at an Orthopedic Hospital in Kolkata. For comparative analysis with the European population, we used results from a previous survey conducted in UK, Spain and Germany.

2.2 Participants

We collected data from 33 users. Figure 1 below shows an age histogram: 10 users were female and 23 male. Nine users were younger but had physical impairments, i.e. cerebral palsy, polio and accidental motor impairment.

We also used results from a previous survey conducted on around 30 people in Spain, the UK and Germany. Details of that survey and demographic detail on users can be found in a separate report (Guide 2011). We used that data to compare the performance and attitude of the Indian population with their European counterparts.

2.3 Functional Parameters Measurement

We measured objective parameters of perceptual, cognitive and motor abilities of users using standard test batteries: minimum legible font size, presence and type of

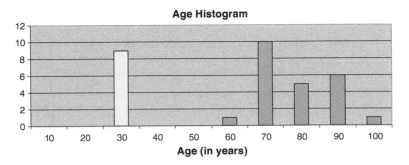

Fig. 1 Age histogram of Indian sample

Education level	Code
Illiterate	0
Pre-school	1
School	2
Graduate	3
Post-graduate	4

Table 1 Coding scheme for education level

colour blindness, grip strength, active range of motion of wrist, scores in trail making test (Army individual test battery 1944) and digit symbol test (Army individual test battery 1944). Detail of these parameters can be found in our previous papers (Biswas and Langdon 2012; Biswas and Robinson 2013).

3 Results

3.1 Cognitive Data Analysis

We analysed the effect of age and education on the cognitive scores of participants. For the Indian population, we found a moderate correlation between age and TMT score ($\rho = 0.38$) and DST Score ($\rho = -0.46$). However, we did not find age to be correlated with TMT and DST scores in the elderly population only, instead, education level significantly correlated with TMT ($\rho = -0.68, p < 0.01$) and DST scores ($\rho = 0.79, p < 0.01$) overall. The graphs used the following coding scheme for education level (Table 1).

We compared the cognitive scores of the Indian population with their EU counterpart. We found both TMT and DST scores are significantly different between EU and Indian samples—European people took less time to complete the Trail Making Task and scored more in the Digit Symbol Task than their Indian

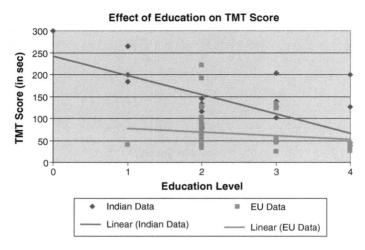

Fig. 2 Comparing effect of education level on TMT score between Indian and EU population

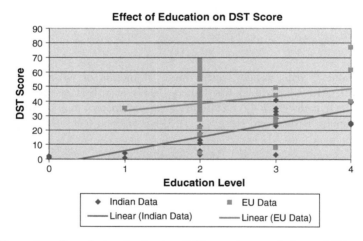

Fig. 3 Comparing effect of education level on DST score between Indian and EU population

counterparts. We also did not find any significant correlation between age, education level and TMT, DST scores for the European population (Figs. 2 and 3).

3.2 Hand Strength Data Analysis

We found that age, gender and height have a significant effect on grip strength for the Indian population. However, we did not find a significant effect of gender and

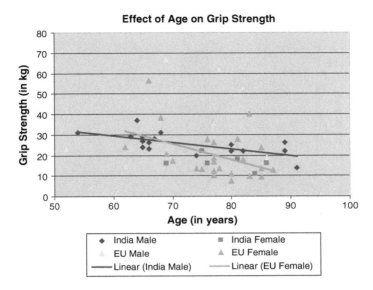

Fig. 4 Comparing effect of age on grip strength between EU and Indian population

height on a similar linear regression analysis on the active range of motion of wrist (ROMW), only age is significantly correlated with ROMW. We compared the hand strength of Indian and European populations. A few European elderly people had grip strength higher than their Indian counterparts, but the ROMW did not differ significantly. We found that for European female elderly people, grip strength correlates moderately with age ($\rho = -0.43$) though age was not found to be correlated with range of motion of wrist (Figs. 4 and 5).

3.3 Visual Data Analysis

Among the Indian population, approximately a third of the sample needed a bigger than standard font size. They could not read the 3 m line in a Snellen chart from 3 m distance. A few users who could read the 3 m line from a 3 m distance commented that they would benefit from a bigger font size on mobile keypads. We found 8 among 33 users could not read plates 16 and 17 of the Ishihara (Colour Blindness 2008) colour blindness test properly: five of them seemed to have protanopia-type colour blindness as they could only read the left hand digit correctly, the remaining three could not read any numbers at all.

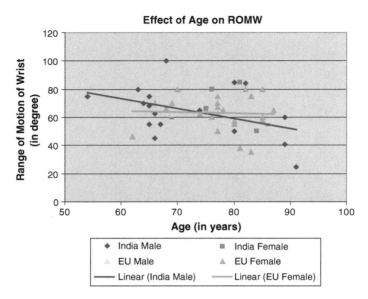

Fig. 5 Comparing effect of age on ROMW between EU and Indian population

3.4 Attitude Towards Technology Data Analysis

Table 2 reports the questions used to understand attitudes towards technology, while Table 3 reports the response of people from different countries. Each question had three possible responses—agree, disagree and neutral.

We did not find any significant difference between elderly people from Spain, the UK, India and Germany regarding their attitude towards technology through a single factor ANOVA. The subjective attitude is discussed in more detail in Sect. 4.

3.5 Technology Exposure Data Analysis

We found that none of the Indian users ever used at computer, Kiosk or smart TV, only a couple used smartphone, while all could use the basic call making and receiving facilities of a mobile phone and could view their favourite television channel. Most of the younger people with physical impairment used computers regularly while three to four elderly users had used computers before though none of the elderly users used one regularly. They reported that they found computers complicated to use, could not remember functionalities and could not find them useful. However, a few wanted to use computers for emailing their sons or daughters or pursuing hobbies like cookery, knitting etc.

Table 2 List of questions used to assess attitude towards technology

1	I think that new technology devices are developed mainly to be used by young users
2	I think that I need to use new technology
3	I consider it important to try to be open-minded towards new technology
4	I consider myself having the necessary skills to manage to use new technology tools
5	I have problems to use these technologies properly even with practice
6	The problems of technology devices are impossible to understand, so it is hard to find a solution
7	When there is a problem with a new technology tool, it is because there is something wrong with that device
8	I'm afraid to touch a new technology tool in case I'll break it
9	I get no advantage from using new technology tools
10	I prefer to use an old-fashioned tool with fewer functions than a new one

Table 3 Response from users from different countries

Question no.	Spain	UK	Germany	India
Percentage of users agreed				
1	30.8	56	46.7	30.43
2	26.9	60	68.8	69.23
3	84.6	72	100	92.31
4	42.3	16	37.5	39.13
5	15.4	56	25	46.15
6	42.3	68	75	15.38
7	15.4	44	31.3	15.38
8	42.3	40	12.5	30.43
9	30.8	54.2	18.8	53.85
10	57.7	68	56.3	52.17
Percentage of users disagreed				
1	69.2	20	40	65.22
2	46.2	28	0	23.08
3	7.7	4	0	0
4	23.1	48	31.3	47.83
5	61.5	28	56.3	30.77
6	30.8	12	12.5	46.15
7	42.3	32	50	53.85
8	38.5	40	68.8	65.22
9	61.5	37.5	68.8	30.77
10	23.1	28	25	47.83

4 Discussion

The aim of this survey was to understand the effects of age on capability and attitudes towards technology, in particular electronic interactive devices like computer, TV, mobile phones and so on. The study produced some interesting insights as detailed below.

4.1 Education Level Is More Important Than Age with Respect to Cognition

We found that age has a moderate correlation with TMT and DST scores even considering younger users, but education level has a significant correlation in retaining task-switching capabilities or retaining visual attention or short-term memory as reflected and measured by the cognitive tests.

Designers should consider the education level of their intended users while designing screen layouts, as a screen with many items may confuse users with lower education levels. This issue becomes more relevant while designing applications for illiterate or neo-literate people in India.

We also found that the average scores in cognitive tests are higher in the European population. One possible reason may be that most European users had a minimum level of school education while that was not true for their Indian counterparts.

4.2 Age Reduces Hand Strength in Turn Capability of Using Computer Peripherals

Our previous study (Biswas et al. 2012; Biwas and Robinson 2013) found that grip strength and active range of motion of wrist is indicative of users' performance with a variety of pointing devices like mouse, trackball, touchscreen and stylus. We found that both Indian and EU populations lose their grip strength with age which negatively impacts their use of standard computer peripherals. The reduced hand strength also makes it difficult to use touchscreens or keypads which was also explicitly mentioned by a few participants.

Designers and interface developers should explore use of hands-free control for elderly users. Eye gaze tracking (Biswas et al. 2013) or brain–computer interfaces (Biswas et al. 2013) can offer valuable alternatives to standard keyboard, mouse and touchscreen interfaces.

Our analysis also developed a model to predict grip strength from the age, height and gender of participants which was reported earlier in the Ergonomics literature (Angst et al. 2010) but not properly validated on the Indian population.

4.3 Bigger Font Size and Buttons Are Good

We found that one-third of users had reduced visual acuity even with a corrective lens. However, most elderly users used glasses and a few of them more than one pair for close and distance vision. It is likely that they may use TV or mobile phones without their glasses or with the wrong glasses on. Their reduced visual acuity compounded with reduced hand strength limits the usability of a standard mobile or remote keypad as well as standard computer or TV interfaces.

Designers should provide bigger buttons and font sizes for elderly people. They should also adjust colour contrast for users with colour blindness.

4.4 Elderly Users Will Use It if It Is Interesting and Easy-to-Use

Most of our elderly users use the basic functionalities of mobile phone and TV as they find them easy-to-use and useful. However, that does not hold true for computers as they find them complicated and not useful. Some elderly users relied on others to send emails to their distant relatives. However, more than 60 % of users in India and Spain believed that new technology was not meant only for young users and more than 80 % of them felt themselves open-minded about using it. More than 60 % of Indian, British and German elderly users wanted to use new technology. However, the verdict on breaking a new tool or using an old-fashioned tool is not very obvious as on average 50 % users agree or disagree while a few have no opinion. Many users especially younger ones with physical impairment emphasised the need for training in learning new technology.

Application developers should think about reducing the complexity of applications intended for elderly users. Applications like email or video chat may be found of interest to elderly users if they are easy to use. Similarly an application deployed on low-end mobile phones or set top boxes has more chance of being acceptable to elderly users as they are already accustomed to those devices.

5 Conclusions

This paper reports results from a survey of capability and attitudes towards technology of elderly users conducted at three different locations in India. It also compares the outcome of a similar survey conducted in Spain, Germany and the UK. The survey finds important results in terms of hand strength and cognitive abilities of users. It proposes a model to predict grip strength from age, sex and height of Indian elderly people, while also reporting a significant effect of education on the scores of cognitive tests. Comparative analysis with European elderly people shows European people tend to sustain better cognitive abilities with age though their attitude towards new technology is not significantly different from their Indian counterparts. Finally it reports subjective views of elderly people about technology and proposes guidelines for future interface designers in developing countries.

References

Angst F, Dreruo S, Werle S, Herren DB, Simmen BR et al (2010) Prediction of grip and key pinch strength in 978 healthy subjects. BMC Musculoskelet Disord 11:94

Army individual test battery (1944) Manual of directions and scoring. D.C. War Department, Adjuvant General's Office, Washington

Biswas P, Joshi R, Chattopadhyay S, Acharya UR, Lim T (2013) Interaction techniques for users with severe motor-impairment. In: Biswas P, Duarte C, Langdon PM, Almeida L, Jung C (eds) Multimodal end-2-end approach to accessible computing. Springer, London

Biswas P, Langdon PM (2012) Developing multimodal adaptation algorithm for mobility impaired users by evaluating their hand strength. Int J Hum Comput Interact 28(9):576–596

Biswas P, Robinson P (2013) A survey on technology exposure and range of abilities of elderly and disabled users in India. In: Proceedings of the 15th international conference on Human Computer Interaction International (HCII), Las Vegas

Biswas P, Robinson P, Langdon PM (2012) Designing inclusive interfaces through user modelling and simulation. Int J Hum Comput Interact 28(1):1–33

Colour Blindness (2008) Ishihara color test. http://www.colour-blindness.com/colour-blindness-tests/ishihara-colour-test-plates/. Accessed on 11 Nov 2013

Dalstra J, Kunst AA, Borrell AE, Breeze C, Cambois E et al (2005) Socioeconomic differences in the prevalence of common chronic diseases: an overview of eight European countries. Int J Epidemiol 34(2):316–326

GUIDE (2011) Initial user tests and model. Document number D7.1. www.guide-project.eu/includes/requestFile.php?id=129&pub=2. Accessed on 7 Nov 2013

ITU (2011) Making television accessible. G3ict-ITU. http://www.itu.int/ITU-D/sis/PwDs/Documents/Making_TV_Accessible-FINAL-WithAltTextInserted.pdf. Accessed on 18 June 2013

Langdon P, Thimbleby H (2010) Inclusion and interaction: designing interaction for inclusive populations. Editor Spec Ed Interact Comput 22:439–448

Mackenbach JP, Kunst AE, Cavelaars AE, Groenhof F, Geurts JJ (1997) Socioeconomic inequalities in morbidity and mortality in Western Europe. Lancet 349(9066):1655–1659

Narasimhan N, Leblois A (2012) Making mobile phones and services accessible for persons with disabilities. A joint report of ITU—The International Telecommunication Union and G3ict—the global initiative for inclusive ICTs

Siegrist J, Marmot M (2004) Health inequalities and the psychosocial environment—two scientific challenges. Soc Sci Med 58(8):1463–1473

User Capabilities Versus Device Task Demands in a Tape Dispenser Product for Persons with Limited Dexterity

Y. M. Choi

1 Introduction

The creation of a product that will be effective in aiding users who have physical limitations presents a designer with many difficult challenges. Users of assistive devices represent a continuum of abilities (Cook et al. 2008), from those with slight to moderate disabilities who may have more general needs to those with more severe disabilities who may have very unique and specific needs. An assistive device may therefore function well for one group of users but poorly for another. This continuum can be a challenge for a company as it often results in small, niche markets for such devices (Cowan and Turner-Smith 1999). It is also a challenge for designers because they draw upon their own experiences and knowledge while investigating potential solutions to a design problem (Norman, 1988). Designers are trained to imagine themselves in the shoes of the user (Nieusma 2004). But since most do not themselves suffer from physical limitations, it is much more difficult to imagine themselves in the shoes of a user who has lost some basic ability to interact with the world that the designer can take for granted. Even if the designer has a physical limitation, unless it is of the same type and severity as the target user's, it may not provide much relevant personal experience to draw upon. Tools or other guidance that can aid designers in eliminating interactions with products that might pose a barrier to use can be important in improving a product's effectiveness as well as an end user's level of satisfaction. This paper will investigate the link between devices designed to aid in the task of taping closed a box and common ergonomic measurements of dexterity.

Upper extremity (UE) function plays an important role in one's ability to perform activities of daily living (ADL) (Desrosiers et al. 1999). UE function may become reduced for a variety of reasons such as age, injury or disease. Accurate

Y. M. Choi (✉)
School of Industrial Design, College of Architecture, Georgia Institute of Technology, Atlanta, GA, USA
e-mail: christina.choi@gatech.edu

P. M. Langdon et al. (eds.), *Inclusive Designing*, DOI: 10.1007/978-3-319-05095-9_2, 13
© Springer International Publishing Switzerland 2014

measurement of UE function is important in evaluating the effectiveness of surgical procedures or rehabilitation strategies. UE function includes motor coordination, manual dexterity, muscle strength, and sensibility (Özcan et al. 2004).

Manual dexterity has been defined as the ability to combine precision and speed with coordinated movements of the arm, hand, and fingers (Özcan et al. 2004). A tool commonly used for testing manual dexterity is the nine hole peg test (NHPT) which is a timed test of finger dexterity and fine motor coordination. It is commonly used by occupational therapists as a quick assessment of dexterity function (Grice et al. 2003; Wang et al. 2011).

Grip and pinch strength are considered to provide an objective measurement of UE integrity (Incel et al. 2002). Grip strength (also known as power grip) is typically measured with a dynamometer where the base rests against the heel of the palm and the handle on the middle of the four fingers. The user then squeezes with maximum effort which is measured by the device. Pinch strength can be measured in a number of ways. One is a "tip pinch" in which the user pinches using the tip of the thumb to the tip of the index finger. Another is the "key pinch" where the thumb pad is pressed to the lateral aspect of the middle phalanx of the index finger. Strength is measured by the user executing the pinches on a pinch meter.

Normative values for various physical abilities can be important resources in helping designers understand the capabilities of target end user. A number of studies have investigated values for grip, pinch, and other measures of UE function. Many have focused on establishing and verifying normative values for healthy adults. Others have studied specific populations, such as children (Poole et al. 2005) or those with loss of dexterity following a stroke (Ada et al. 1996).

In order to design products where the required user interaction does not present a barrier to use, it is important to understand which particular aspects of dexterity are most important in the performance of a desired task. This can allow the product to be designed so that it does not require certain physical actions and/or provides a substitute for required actions that the end user might not be able to perform. While normative data for various measures of dexterity exist, these relationships are not known which can limit their utility to a designer as a tool to avoid barriers to use. This paper investigates the relative values of the UE functions in eight products designed for a specific task.

2 Method

The data presented here were collected as part of a study to investigate the effect that different types of input provided during the course of designing a new product have on an end user's opinion of the usability of the finished prototype (Choi 2011). Eight new devices were designed (each by an independent design team) with the single goal of aiding a user with limited dexterity to perform the task of taping a box closed. Each team received the same design brief outlining the product use scenario and information about dexterity. The only difference between teams was the additional information they were allowed to gather and use (some were allowed no additional data, some were allowed only textual research, some were allowed only simulation and some were allowed only input from users). Twenty people with limited dexterity were recruited to evaluate the resulting devices. Limited dexterity was defined as reduced ability in grasping, holding, squeezing, or fine finger manipulation. Participants were recruited primarily from the CATEA Consumer Network (Choi et al. 2008) and other disability specific organizations in Atlanta, Georgia. Participants were eligible if they were aged 18 years or older, had a dexterity limitation and did not have a sensory limitation (i.e., hearing or vision).

Each potential participant was prescreened using the DASH survey: disabilities of the arm, shoulder, and hand (IWH 2009). The DASH was developed to provide a quick, self-administered measure of symptoms and functional status (Hudak et al. 1996). It is a 30 item survey that asks the taker to rate their ability to perform various activities of daily living. Scores can range from 0 (a person has no difficulties in performing any of the tasks) to 100 (a person is unable to perform any of the tasks at all). The scores were used to help ensure that participants represented a wide range of functional abilities.

Before evaluating the devices, measurements were taken of each participant's grip strength, key pinch strength, tip pinch strength, and performance on the NHPT. Three measurements of the grip, key pinch, and tip pinch strength were taken for each hand using a Jamar dynamometer and averaged for each hand. The NHPT was performed twice by each participant: once with the left and once with the right hand. Each participant identified their dominant hand during recruitment.

After performing the dexterity tests, each participant evaluated each of the eight new devices. The evaluation consisted of using each to perform the task of taping closed an identical box which was pre-loaded and placed next to it. The task was timed from the moment the evaluator touched the device to begin the task to the moment that sealing the box was completed or the participant gave up. A standard gun style tape dispenser was always evaluated first to provide a point of reference for how well the studied devices performed in comparison. The studied devices were then evaluated in a randomly assigned order.

A usability survey was completed immediately after the task was finished or attempted with each device. The survey itself is reported in (Choi 2011). It is a Likert item survey designed to capture the evaluator's opinion about the level of effectiveness, satisfaction, safety, ease of use, comfort, and other major components of usability. It was based on the existing USE Questionnaire (Lund 2001) and System Usability Scale (Bangor et al. 2008) and validated prior to the study. At the very end, participants were asked to rank each device in order from best to worst based on their opinion. Each device was given a name for internal reference (it was not used with participants) based roughly on how it was setup or operated.

The pole device (Fig. 1a) is made up of a "shelf" that travels up and down a pole and can be rotated to the left or right. The shelf contains a pair of "arms" that hold down the flaps on a box. The arms are powered and can be extended and

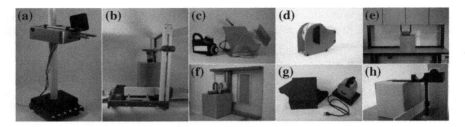

Fig. 1 The devices used in the evaluation. **a** The pole, **b** the conveyor, **c** the feeder, **d** the puller, **e** the cabinet, **f** the wall, **g** the stamper, and **h** the pole device

retracted as needed. A tape dispenser unit is located on the top of the shelf and is attached to manual telescoping arms. A cutting blade located on the bottom corner of the tape dispenser cuts the tape when the dispenser is rotated. All of the powered operations can be controlled from foot buttons on the base of the device. The device rolls on locking casters so that the user can roll it into position, adjust the taping shelf using the foot controls, hold down the box flaps by extending the arms and lowering the shelf. Tape is dispensed to the top by affixing the exposed part of the tape to the side of the box and pulling the dispenser over the top. Operation requires the use of pinch or key grip to attach the tape and power grip for operating the dispenser.

The conveyor device (Fig. 1b) allows a loaded box to be placed on a small conveyor belt which pulls the box under a tape dispenser mounted on a horizontal platform. This can be moved up and down by using a blue-colored handle that is integrated into the vertical support. Guides on the side of the conveyor can be adjusted to the width of the box via a second blue-colored handle. The tape is affixed to the leading edge of the box and the belt pulls the box under the dispenser causing tape to be applied to the center. A spring mechanism on the dispenser causes a blade to cut the tape once the box has passed through the device. Operation requires the use of pinch or key grip to attach the tape and power grip for adjusting the dispenser height.

The feeder device (Fig. 1c) is designed to sit on the edge of a table and dispense tape in 3-inch increments by pressing a green "Dispense" button on top of the unit. When the desired length of tape has been dispensed, it is cut free by pressing a red outlined "Cut" button on the front of the unit. The dispenser has a small lip that extends down on the front of the device. This provides an opening for the tape to come out and also prevents the unit from sliding away from the user when the dispense button is pressed. After the tape is cut, the user can then place it on the desired area of the box. Operation requires the use of pinch or key grip to grasp the tape and place it on the box.

The puller device (Fig. 1d) is operated by inserting one hand into the central hand hold. The end of the tape is then attached to the edge of the box. The dispenser is then pulled across the top of the box to dispense the tape. There are a series of rollers around the outside of the dispenser to help make the dispensing of the tape more smooth. A cutting blade is attached behind the last roller that cuts the tape with a downward motion to bring the blade in contact with the tape. Operation requires the use of pinch or key grip to attach the tape and power grip for operating the dispenser.

The cabinet device (Fig. 1e) is a dispenser unit mounted under a cabinet or shelves. Telescoping arms allow the height to be adjusted and tracks under the cabinet allow the front-to-back position to be adjusted. Tape is applied to the box by sliding it across the table surface under the dispenser. A curved plastic guide presses the tape down onto the box as it passes under. Pulling downward causes a blade attached to the rear edge of the plastic guide to cut the tape after the box has passed under the dispenser. Operation requires the use of pinch or key grip to attach the tape and power grip for aligning the dispenser.

The wall device (Fig. 1f) consists of a tape dispenser which is mounted on tracks that allow the dispenser to be moved up, down, forward, or backward to align the tape with a box. The leading tape edge is affixed to the box and the dispenser is pushed over the top. When the dispenser reaches the back edge of the box, it naturally slides down, which brings a blade in contact with the tape to cut it. Operation requires the use of pinch or key grip to attach the tape and power grip for operating the dispenser.

The stamper device (Fig. 1g) is a handheld device with a Velcro strap and button on the top that can be used with one hand. To use, the stamper is placed directly on the location where tape should be dispensed. The tape is then "stamped" onto the box when the button on the top is pressed. Operation is not dependent on power grip, key, or tip pinch.

The table device (Fig. 1h) is designed with a tape dispenser component attached to a telescoping pole mounted to a table or other workspace. A dowel extending from the dispenser holds down the box flap on the far side of the user. A low cutout on the dispenser allows a user to put a hand under the tape, sticking it to the top of the hand, and then lift and pull the tape across the top of the box. The user's arm will naturally hold down the box flap nearest to them. Affixing the tape to the far edge of the box with a downward motion also unsticks the tape from their hand. The strip of tape is pressed down against the box, starting on the side farthest from the dispenser. The tension created causes the tape to be cut by a small blade on the top edge of the dispenser. Operation requires the use of pinch or key grip to attach the tape.

3 Results

The dexterity measurements were correlated with the usability scores, the task times, and the user's product rank for each of the devices across all study participants. Data collected for task time (s), NHPT (s), grip (kg), key pinch (kg), and tip pinch (kg) are continuous data. Data collected for the usability score and product rank are ordinal data. Pearson's (r) correlation was used for comparisons

Table 1 Pearson's (r) correlation where $p < 0.05$ between task time for each product and participant dexterity measurements

Task time (r)

Measurement	Cabinet	Wall	Stamper	Pole	Feeder	Control
NHPT-D	0.65	0.46	0.51	−0.53	0.49	0.52
NHPT-ND	0.69				0.67	

The conveyor, table, and puller products showed no significant correlations

Table 2 Spearman's rho (r_s) correlation where $p < 0.05$ between product rank for each product and participant dexterity measurements

Product rank (r_s)

Measurement	Conveyor	Wall	Table
DASH	−0.54		
NHPT-D	−0.55		
NHPT-ND	−0.54		
Grip-ND			−0.5
Tip-ND			−0.46
Key-D		−0.5	

The cabinet, stamper, puller, pole, and feeder products showed no significant correlations

involving all continuous data. Spearman's rho (r_s) correlation was used for comparisons that included strictly ordinal data.

No significant correlations were found between the usability score and the dexterity measurements. Significant correlations ($p < 0.05$) were found between the measured task times and some of the dexterity measures as shown in Table 1. Some significant correlations were also found between the product rank and the dexterity measures as shown in Table 2.

A number of study participants had mobility limitations along with their dexterity limitation. Thus, the data were also divided into dexterity only and dexterity+mobility groups. Correlations between the dexterity measures and usability, task time, and rank were examined for each group. Tables 3, 4, 5, 6, 7, and 8 show the significant correlations ($p < 0.05$, $p < 0.001$ indicated by **) that were found.

4 Discussion

The most consistent overall correlation between physical ability and the devices in this study was between the NHPT measurement of the user's dominant hand, and the task duration. Intuitively this makes sense. Longer times for the NHPT indicate less dexterity. A positive correlation might generally be expected since a lower level of dexterity should cause the task to take longer to perform. The pole product

Table 3 Spearman's rho (r_s) correlation where $p < 0.05$ between usability score for each product correlated and participant dexterity measurements for non-wheelchair participants

Usability score (r_s)

Measurement	Wall	Table	Pole
NHPT-ND		0.58	0.6
Tip-ND	−0.64		

The conveyor, cabinet, stamper, puller, feeder, and control products showed no significant correlations

Table 4 Pearson's (r) correlation where $p < 0.05$ between task time for each product and participant dexterity measurements for non-wheelchair participants

Task time (r)

Measurement	Cabinet	Wall	Table	Pole	Control
DASH	0.68		0.65	0.76**	
NHPT-D		0.66		0.71**	0.52
NHPT-ND		0.52			0.67**

$p < 0.001$ is indicated by **. The conveyor, stamper, puller and feeder showed no significant correlations

Table 5 Spearman's rho (r_s) correlation where $p < 0.05$ between product rank for each product and participant dexterity measurements for non-wheelchair participants

Product rank (r_s)

Measurement	Puller	Pole
NHPT-D		−0.6
Tip-D	−0.77	

The conveyor, cabinet, wall, table, stamper, feeder, and control products showed no significant correlations

Table 6 Spearman's rho (r_s) correlation where $p < 0.05$ between usability score for each product and participant dexterity measurements for wheelchair participants

Usability score (r_s)

Measurement	Wall	Stamper	Puller
NHPT-D			−0.76
Grip-ND	0.74		
Key-ND		−0.72	

The conveyor, cabinet, table, pole, feeder, and control products showed no significant correlations

was an exception with a negative correlation, suggesting that users with better dexterity actually have a more difficult time using it.

Though each of the products in this study was designed specifically with dexterity limited end-users in mind, more correlation between the physical

Table 7 Pearson's (r) correlation where $p < 0.05$ between task time for each product and participant dexterity measurements for wheelchair participants

Task time (r)			
Measurement	Conveyor	Wall	Control
NHPT-D			0.52
NHPT-ND	0.72	0.72	0.67
Grip-ND		0.74	
Key-D	0.75		
Key-ND		0.71	

The cabinet, wall, table, puller, pole, and feeder products showed no significant correlations

Table 8 Spearman's rho (r_s) correlation where $p < 0.05$ between product rank for each product and participant dexterity measurements for wheelchair participants

Product rank (r_s)		
Measurement	Conveyor	Stamper
NHPT-D	−0.82	
NHPT-ND		−0.73

The cabinet, table, puller, pole, feeder, and control products showed no significant correlations

measurements of dexterity and the evaluations of the products was expected. From a designer's perspective it would be valuable to know what kind of an effect a particular component of dexterity such as key pinch has on the usability, efficiency, or user preference for a product. It would have been especially helpful to find correlations among grip, key pinch, and tip pinch strength as normative values are available for these measurements. Even though norms are based on measurements from populations without any specific limitations, they can provide a baseline guide of reasonable expectations. So if it were found that key pinch is especially important to the task of taping a box, a designer can begin with the normative expectation for key pinch force and reduce it so that any key pinch force required in a product is at an appropriate level for the target users.

Correlations with many of the standard dexterity measures did not emerge in this study. This could be due to a number of reasons. It is possible that each of the eight design teams did a good job at removing any significant need to use grip, key pinch, or tip pinch. This explanation seems unlikely due to the fact that the standard tape dispenser did not show any correlation with these traits (it at least requires a power grip) nor did the new devices show consistent correlation.

Several of the participants had other limitations. The most common were mobility limitations. Eight participants were confined to a wheelchair (either manual or powered). In some cases, this could have a major impact on the way that the user interacted with and actually attempted to use the product. The performance of wheelchair and non-wheelchair users was analysed separately. Even in this analysis, there is no consistent dependency among grip, key, or tip pinch strength for either wheelchair or non-wheelchair users performing the tasks.

The best measure that correlated with the task performance (efficiency) was the NHPT, particularly the NHPT score for the dominant hand. As opposed to the grip, tip, and key pinch measures, better scores on the NHPT (i.e., faster times) require aspects of upper body movement as well as gross and fine motor control of the arm, hand, and fingers. Similarly, the DASH measure was correlated with the task time and usability scores of some products. It is also a broader measure of capabilities, although the score is compiled through user self-reporting rather than through an objective test.

The small dataset in this study limits the generalisability of any conclusions. Still the results indicate the possibility that broader measures of ability might be more useful as design aid to predict performance and usability than more atomic measurements such as grip or pinch strength. Even though broad measures like the DASH and NHPT are used often, normative values are not commonly available for large, representative groups of users. Using only these, a designer would not have an easily available reference of the typical capabilities of a particular subset of users. This can be gathered manually just as with more common measures, though collecting a large and representative sample is very time consuming and can be expensive, which makes it less practical. Also, since the DASH and NHPT measures aggregate multiple abilities, the designer is still left with the problem of not knowing which component of a user's ability might impact their use of a product. The specific way that a task is accomplished by a device is obviously greatly influenced by how the interface/interaction with the user is designed. As evidenced by the products made by the design teams in this study, the same task outcome can be achieved in a vast array of completely different ways. Even with a DASH or NHPT score, without understanding how the individual components that make up the score might influence interaction with a product, they are not very useful in identifying barriers that a user might encounter in a designed product.

5 Conclusion

Commonly available ergonomic measurements such as grip, key pinch, and tip pinch strength, while clearly components of dexterity, do not alone appear to be reliable predictors of product design outcomes such as usability or task efficiency. Even for a specific task such as taping a box closed, the user's interaction with the product involves many other factors that these individual measurements do not capture. More broad measures of dexterity, such as the DASH and NHPT, both appear to be much better at predicting how well a user might or might not be able to perform a given task with a device. These broader measures might be useful for evaluating the performance of a product for a group of users after it has been designed. They are more limited as a design tool since they are not typically available for broad and representative groups of users. They also cannot inform a designer on whether individual aspects of the physical interaction with a product might pose a barrier to use because they aggregate many types of interactions into a single measurement.

Further study would need to be done to better understand the role that different components of physical ability play in the performance of specific tasks. It may be found that certain abilities generally affect a broad group of tasks or are specific to particular ones. There are other ergonomic measurements, such as wrist strength, that can affect dexterity that were not considered in this study. It would be useful to collect more physical measurements for similar studies in the future, since even a very simple task can involve a much larger number of physical movements. This may help identify already available ergonomic measurements (or combinations of measurements) that might serve as useful design aids for a given task.

Perhaps some different types of measurements could serve as useful aids. As the relationships between particular physical actions and the impact on associated tasks are determined, more effective ways of measuring them might be derived. As the DASH measurement shows correlations with the product performance in this study, it may be possible to devise a reliable way to identify attributes that pose use of barriers through self-reporting, rather than direct physical measurement, allowing relevant data to be collected more quickly, widely, and inexpensively.

References

Ada L, O'Dwyer N, Green J, Yeo W, Neilson P (1996) The nature of the loss of strength and dexterity in the upper limb following stroke. Hum Mov Sci 15(5):671–687

Bangor A, Kortum P, Miller J (2008) An empirical evaluation of the system usability scale. Int J Hum Comput Interact 24(6):574–594

Choi YM (2011) Managing input during assistive technology product design. Assist Technol 23(2):65–75

Choi YM, Sabata D, Todd R, Sprigle S (2008) Building a consumer network to engage users with disabilities. In: Langdon P, Callahan J, Robinson P (eds) Designing inclusive futures. Springer, London

Cook A, Polgar J, Hussey S (2008) Assistive technologies: principles and practice, 3rd edn. Mosby, St. Louis

Cowan D, Turner-Smith A (1999) The role of assistive technology in alternative models of care for older people. Research, HMSO 2:325–346

Desrosiers J, Hébert R, Bravo G, Rochette A (1999) Age-related changes in upper extremity performance of elderly people: A longitudinal study. Exp Gerontol 34(3):393–405

Grice KO, Vogel KA, Le V, Mitchell A, Muniz S et al (2003) Adult norms for a commercially available nine hole peg test for finger dexterity. Am J Occup Ther 57(5):3

Hudak PL, Amadio PC, Bombardier C (1996) Development of an upper extremity outcome measure: The DASH (disabilities of the arm, shoulder, and head). Am J Ind Med 29:6

Incel N, Ceceli E, Durukan P, Erdem H (2002) Grip strength: Effect of hand dominance. Singap Med J 43(5):602–608

IWH (2009) DASH and QuickDASH: conditions of use. www.dash.iwh.on.ca/conditions.htm. Accessed on 24 Oct 2013

Lund AM (2001) Measuring usability with the USE questionnaire. Usability User Exp 8(2):8

Nieusma D (2004) Alternative design scholarship: working toward appropriate design. Des Issues 20(3):13–24

Norman D (1988) The design of everyday things. Basic Books, New York

Özcan A, Tulum Z, Pınar L (2004) Comparison of pressure pain threshold, grip strength, dexterity and touch pressure of dominant and non-dominant hands within and between right- and left-handed subjects. J Korean Med Sci 19(6):874–878

Poole J, Burtner P, Torres T, McMullen CK, Markham A et al (2005) Measuring dexterity in children using the nine-hole peg test. J Hand Ther 18(3):348–351

Wang Y-C, Magasi SR, Bohannon RW, Reuben DB, McCreath HE et al (2011) Assessing dexterity function: A comparison of two alternatives for the NIH toolbox. J Hand Ther 24(4):213–321

Part II
Designing Cognitive Interaction with Emerging Technologies

Three Scanning Methods for Text Cursor Manipulation

A. J. Sporka, O. Polacek and J. Murin

1 Introduction

The set of keys dedicated solely to cursor movement control evolved over decades. A standard PC has four arrow keys: Home, End, Page Up and Page Down whose function is well-known to the vast majority of computer users. Jumping to the beginning or end of the document or moving left or right by one word is usually possible using the Control key as a modifier. By repeated pressing of these keys, the users may carry out various cursor movement tasks.

It is a problem for people with reduced dexterity in their hands to use the standard keyboard and mouse. Either their performance is limited or they are entirely prevented from working with these devices. A number of assistive techniques have been developed that enable users to type text on a standard keyboard or via dedicated hardware devices.

A specific class of techniques is based on using a single switch (key). They are intended for people with severe motor impairments. The switch can be operated for example by chin (www.hawking.org.uk/the-computer.html) or by raising an eyebrow (Felzer et al. 2006). One-switch interfaces are frequently based on scanning. A limited number of options is presented sequentially, one by one. The option can be selected by pressing the switch only when active.

A number of one-switch techniques exist that are intended to emulate the process of typing and text editing. Most literature about scanning keyboards focuses on character entry. Nevertheless, as reported by Whiteside et al. (1982), as many as 25 % of keystrokes are made with the intention of moving the cursor. Text cursor movement is however not often the centre of attention of present assistive technology.

A. J. Sporka (✉) · O. Polacek · J. Murin
Department of Computer Graphics and Interaction, Faculty of Electrical Engineering,
Czech Technical University in Prague, Prague, Czech Republic
e-mail: sporkaa@fel.cvut.cz

P. M. Langdon et al. (eds.), *Inclusive Designing*, DOI: 10.1007/978-3-319-05095-9_3,
© Springer International Publishing Switzerland 2014

The aim of this paper is to describe three approaches to one-switch control of text cursor movement and present a comparison performed by a sample of able-bodied participants.

2 Scanning Input and Control

Scanning systems and keyboards have been studied extensively in past decades. Scanning refers to a selection technique in which a number of items are highlighted sequentially until one item is selected and the corresponding command is executed. Scanning is based on two atomic operations: *step* and *selection*. The step operation highlights the next item in a predefined order, while the selection operation executes a command assigned to the highlighted item.

This technique requires only one signal from the user (further referred to as switch) which is mapped either to step or to selection. The other operation is then triggered after a predefined scanning interval is reached. Based on the mapping, we can distinguish among several scanning modes. The most prevalent are described below:

- Automatic scanning is the most common. The selection is controlled by the users and a step is triggered automatically after a scanning interval expires.
- Step scanning (Miro-Borras et al. 2008) is similar to automatic scanning, but the control of selections and steps is reversed: steps are controlled by the user and selections are automatic.
- Self-paced scanning (Felzer et al. 2008) distinguishes single and double switch activations. The single switch activation is used as a scan step and double switch activation as a scan selection.
- Inverse scanning (Miro-Borras et al. 2008) requires a switch with two states. When the switch is activated, the automatic scanning starts. Once released, the selection is done by waiting for the scanning interval.

When two switches are available, the scanning interval is usually replaced by the second switch. A number of devices or interaction modalities can be used as a switch.

Several scanning techniques exist that have been used for text entry as well as for menu selections or browsing the contents of a menu. In linear scanning, e.g. (Schadle 2004), items are sequentially highlighted in one group until the correct one is selected. Row–column scanning (Koester et al. 1994) improves selection for a greater number of items by organising them in a matrix. Selection of an item is done in two levels. Three-dimensional scanning (or group scanning) (Lin et al. 2008; Felzer et al. 2009) reduces the number of scanning steps by adding one more level. Binary scanning (or dual scanning) (Harbusch et al. 2003; Baljko et al. 2006;

Felzer et al. 2008) recursively splits items into two halves until a single item is highlighted. *n*-ary scanning (Polacek et al. 2012) is the generalisation of the binary scanning. Ternary scanning was found to be optimal among other *n*-ary scanning techniques for the character input.

Controlling a text cursor is essential for editing longer texts. However, previous works focus mainly on character input. Only two papers were found reporting on the involvement of cursor control for motor-impaired people. Felzer et al. (2006) presents a text editor operated by intentional muscle contractions. Their editor includes emulation of arrows, PgUp, PgDn, Home and End keys. Wandmacher et al. (2008) describes an assistive communication system which includes a Navigation Keypad for cursor control. The whole system is operated by scanning. Neither paper, however, reports a study focusing specifically on cursor control.

3 Method Designs and Prototypes

Designated keys on a standard PC keyboard enable the users to move the cursor by a selected unit of text: character, word, line and page. By pressing a key the user chooses the unit of movement, chooses the direction and issues a command to perform the movement.

For the purposes of our study we have designed three methods that emulate this process by a single switch, which is triggered by the space bar in our prototype. The interaction in all methods is carried out cyclically. Each cycle can be divided into these stages: (1) unit selection, (2) direction selection and (3) performance. The methods differ in the implementation of these stages. They are described in the following paragraphs.

Step scanning method. The user interface of our prototype of this method is shown in Fig. 1. It consists of three parts: the leftmost part is document content (chapter names and numbers) and the rightmost part is a menu in which the user selects a command for cursor movement. The document is displayed in the middle.

The state diagram of this method is shown in Fig. 2a. The interface gives the user a choice of unit of movement. Upon pressing a key the interface enters the direction selection stage. The travel can be either towards the beginning or the end of the document, by one selected unit of movement at a time.

The cursor moves immediately upon selecting the direction (third stage). The cursor can be moved again by the same unit in the same direction by repeated activation of the switch unless a timeout is encountered (the duration of the scan interval). In such case, the interface returns to the second stage (direction selection) and the user may select the opposite direction of movement (useful in case of target overshoot). If another timeout is encountered, the interface returns to the unit selection stage.

Automatic scanning method. The layout of the user interface is shared with the step scanning method (see the prototype in Fig. 1). The first two stages are also

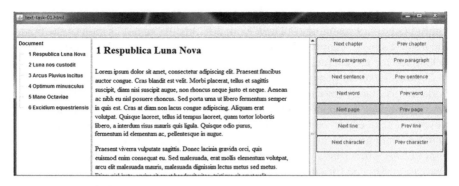

Fig. 1 User interface of the prototypes

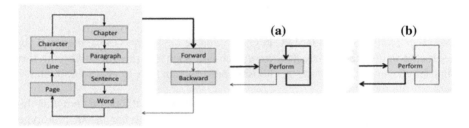

Fig. 2 a Step and **b** automatic scanning methods. *Thin arrows*—scan steps (scan interval expiration), *thick arrows*—scan selections (switch activated by the user)

Fig. 3 Hierarchical scanning method: Selecting words within a sentence. The lines in the figure represent individual states of the user interface, as visible during interaction

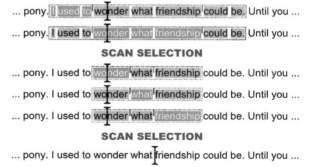

shared: First, the user selects the unit and the direction of movement. The performance is carried out automatically upon the scanning interval expiration and henceforth repeatedly with the same interval (see Fig. 2b). The motion performance continues until stopped by activating the switch.

Hierarchical scanning method. This method shares the movement unit selection stage with the other methods. However, there is no explicit selection of the direction of the movement.

Available units of movement are chapters of the current document, paragraphs of the current chapter, lines of the current paragraph, words of the current line or characters of the current word. An example of operation is shown in Fig. 3. In this example, the user chooses to select words within the current sentence. When a word is selected, the cursor moves to its beginning.

In order to reduce the total number of scan steps, the units are organised in a ternary tree. The interface displays the entire range of text from which it is possible to select (grey background) as well as the highlighted item (red).

The first two methods were designed as a direct application of scanning techniques for keyboard emulation whereby individual command selections correspond to cursor key strokes which results in moving the cursor by one movement unit (character, line, etc.). In contrast to this, the hierarchical scanning method is based on an absolute specification of the position within text.

4 User Study

The performance of all three methods was compared in an experiment with 15 able-bodied participants (one female, 14 male, mean age = 24.1, std. dev = 4.37), recruited from the CTU students. All of them were experienced computer users with previous experience with a text editor. They did not receive any compensation for their participation.

Each participant completed a single session which consisted of demographic questionnaire, two testing blocks (training and measurement) and post-test questionnaire. In one block, they completed 13 tasks with all three methods described above. In each task, they were asked to place the text cursor to a defined location in text (a technical article, approx. 3800 words of length) depicted in the text by a red circle with yellow background. The chapter, which contained the target, was highlighted in the overview of chapters. The order of tasks was identical for all methods and participants. The order of methods was counterbalanced using Latin square to compensate for a learning effect. A typical session lasted approximately 1 hour. The experiment was designed as a within-subject 3×13 experiment with two factors: three methods and 13 tasks. The dependent variables were:

- Task time. The task time was measured starting with the first selection until the cursor was set on the target.
- Scan step count. Number of scan steps in each task.
- Scan selection count. Number of scan selections in each task.

- Selections per scan steps (SPS). According to (Mackenzie et al. 2010), this metric captures the cognitive or motor demand on users. It is defined by Eq. (1). With high SPS number the interaction is more demanding as the user does not have enough time to rest during the passive scan steps and to think over the selection strategy.
- Scanning efficiency (SE). The user does not always choose the optimal path or the optimal choice of commands. Thus, we use SE metric, which captures how optimally each method was used (see Eq. 2).

$$\text{SPS} = \frac{\text{selections}_{\text{user}}}{\text{steps}_{\text{user}}} \tag{1}$$

$$\text{SE} = \frac{\text{selections}_{\text{user}} + \text{steps}_{\text{user}}}{\text{selections}_{\text{optimal}} + \text{steps}_{\text{optimal}}} \tag{2}$$

4.1 Objective Results

A two-factor ANOVA showed significant interaction between task and method for all dependent variables, namely, time ($F(24, 546) = 5.258$, $p < 0.01$), scan steps ($F(24, 546) = 7.663$, $p < 0.01$), scan selections ($F(24, 546) = 7.037$, $p < 0.01$), selections per scan step ($F(24, 500) = 18.538$, $p < 0.01$), SE ($F(24, 546) = 3.636$, $p < 0.01$) and relative effort ($F(24, 546) = 5.807$, $p < 0.01$). Thus, we analysed the results of all dependent variables separately for each task using ANOVA and Tukey's HSD for post hoc comparisons. Conclusions from these tests are presented in following paragraphs.

Means and standard deviations for all measured variables are shown in Fig. 4. The values are shown separately for each task. Note that the tasks are ordered according to the distance between the start and end position of the cursor in the text. For example, carrying out Task 9 (T9) required travelling the shortest distance with the cursor.

Task time. Step scanning was the fastest method in almost all tasks. It was significantly faster than automatic scanning in all tasks. The hierarchical scanning was mostly faster than automatic and slower than step scanning. Exceptions to this are three shortest-distance tasks (T9, T3, T5) where the hierarchical scanning was the slowest method. Automatic scanning was found the slowest in all other tasks.

Scan steps. The number of scan steps was significantly the lowest in the case of step scanning for each task. The number of scan steps used in hierarchical scanning is significantly the highest in three shortest-distance tasks (T9, T3, T5) but it is significantly lower than automatic scanning in the longest task (T8).

Scan selections. The number of scan selections of automatic scanning is the lowest in most tasks. It is always significantly lower than step scanning.

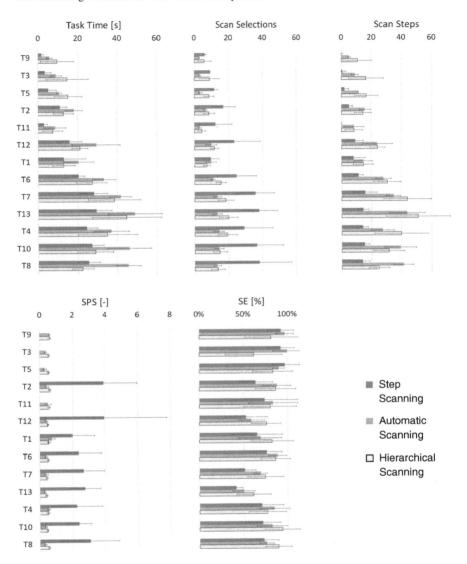

Fig. 4 Measured variables in the experiment. Tasks are ordered by target distance

Hierarchical scanning significantly is lower than step scanning in most tasks except T9 and T3. It is not significantly higher than automatic scanning in a total of eight tasks.

The step scanning uses scan selections to move the cursor in the text, while automatic scanning uses scan steps. Since selections have to be shorter than scan steps, the step scanning is faster. The hierarchical scanning reduces number of scan selections in comparison to the automatic scanning which becomes even more evident in tasks with longer target distance. The hierarchical scanning is thus faster than the automatic scanning except for short-distance tasks.

Selections per scan step (SPS). The SPS rate for the step scanning was significantly higher than automatic and hierarchical scanning for all tasks except four (T9, T3, T5, T11). No scan step in the step scanning was required in these tasks, thus, the SPS rate in this case is undefined. No significant difference was found between automatic and hierarchical scanning. The grand mean (the mean among all tasks) for step scanning was 2.82 (SD = 1.93) after excluding the four tasks. Grand means for automatic and hierarchical scanning were 0.44 (SD = 0.15) and 0.52 (SD = 0.09) respectively.

The high speed of the step scanning is outweighed by a high SPS rate which makes this method significantly more demanding on users. This might be problematic for switch modalities which require high user effort.

Scanning efficiency (SE). The hierarchical scanning was significantly more efficient than step scanning in five tasks (T2, T7, T13, T8, T12) and a automatic scanning in three tasks (T13, T8, T12). Automatic scanning was significantly more efficient than step scanning in two tasks (T2, T7). Grand means for step, automatic and hierarchical scanning were 71.2 % (SD = 26 %), 80.1 % (SD = 21 %) and 79.7 % (SD = 24 %), respectively. The grand mean for step scanning is the lowest. This implies that participants did not always choose the optimal path of the text cursor. In automatic and hierarchical scanning, participants had more time to think of cursor movement strategies thanks to more scan steps.

Usability aspects. Switching from the menu to text and vice versa while scanning had a negative impact on performance. The users had to switch rapidly between two foci of attention—the text cursor and the highlighted item in the menu. This problem was common for all methods. Two other problems were found in the hierarchical scanning:

- *Missing Undo* This problem became evident, when participants selected the wrong unit of movement of scanning (e.g. chapters instead of letters). In this case, they had to repeat scan through all units of movement in the hierarchy, which significantly slowed them down.
- *Long text* The performance in T13 was affected by the fact that the target was not visible on the screen due to the chapter length while scanning paragraphs. This can be easily solved by employing n-ary scanning with higher n so that a highlighted item always fits on the screen.

4.2 Subjective Results

The acceptance of the three methods was evaluated by a simple questionnaire administered to each participant at the end of the session. The questionnaire was designed to measure four aspects of the user experience for each method by the following 5-point Likert items: speed 'Method X was fast', accuracy 'Method X

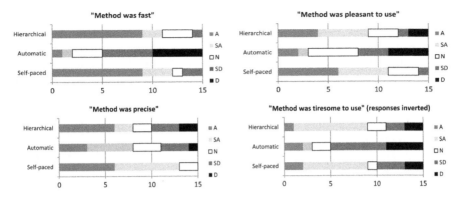

Fig. 5 Subjective evaluation responses

was precise', comfort 'Method X was pleasant to use' and fatigue 'Method X was tireing to use'.

The responses were 'agree', 'somewhat agree', 'neutral', 'somewhat disagree' and 'disagree'. (The responses on fatigue were inverse coded for the purposes of evaluation). All responses are shown in Fig. 5.

We performed a Friedman test for each of the four aspects. A significant difference was found for speed ($\chi^2(2, N = 15) = 14.27$, $p < 0.001$) and comfort (χ^2 (2, $N = 15$) $= 7.26$, $p < 0.05$). A trend was found for fatigue ($\chi^2(2, N = 15) = 5.06$, $p < 0.1$).

A pairwise post hoc testing was performed using a series of Wilcoxon rank sum tests for speed and comfort. Automatic scanning performed significantly worse in speed than other two methods, the step scanning ($W(2) = 27.5$, $p = 0.0003$) and the hierarchical scanning ($W(2) = 199.5$, $p = 0.0002$). Step scanning was evaluated significantly better for comfort than automatic scanning ($W(2) = 44.5$, $p = 0.0041$).

5 Conclusion

Three methods of text cursor control have been described in this paper: step scanning, automatic scanning and hierarchical scanning. This paper also reports on a user study whose aim was to compare the performance of these methods.

The step scanning method was found to be the fastest method of the three. It requires the lowest number of scan steps (less waiting). At the same time it requires a high number of scan selections (more frequent activation of the switch), and therefore it is suitable for switches that are not demanding in terms of physical effort. The step scanning was found more comfortable by the users than the automatic scanning.

The automatic scanning, which uses scan steps to move the cursor, was found the slowest of the three. It requires a low number of scan selections and a high number of scan steps. The selections per scan step rate are relatively low which makes the method usable for modalities which require high user effort. The perceived speed of the automatic scanning was also the lowest of all three methods.

The hierarchical scanning was found suitable for cursor movements over longer distances. This method is faster than the automatic scanning but slower than the step scanning. The number of scan steps and selections of this method are comparable to the automatic scanning which results in small amount of selections per scan step. The SE is also comparable to the automatic scanning. The method shows a high relative effort for cursor travels over short distances.

The step scanning is optimal for people who can use a rapid and comfortable modality to control the switch. If this is not the case, a combination of automatic and hierarchical scanning seems optimal. For longer distances, hierarchical scanning can be used as it almost reaches the speed of the step scanning while saving the physical effort. Cursor movements over shorter distances can be done with automatic scanning, which does not perform significantly worse than step scanning over those distances.

The caveat of this study is that able-bodied participants were chosen instead of the intended target group. A formal study with people with motor disabilities will be performed as a follow-up.

We also plan to further investigate the efficiency of different sets of commands for cursor manipulation and their arrangement on screen. We also need to find the optimal n for the n-ary scanning for different units of movement in the document.

Acknowledgments This study was partially supported by project TextAble (H12070; Ministry of Education, Youth and Sports of the Czech Republic.)

References

Baljko M, Tam A (2006) Indirect text entry using one or two keys. In: Proceedings of the 8th international ACM SIGACCESS conference on computers and accessibility, Portland, US

Felzer T, Nordmann R (2006) Speeding up hands-free text entry. In: Proceedings of CWUAAT'06, Cambridge, UK

Felzer T, Rinderknecht S (2009) 3dScan: an environment control system supporting persons with severe motor impairments. In: Proceedings of the 11th International ACM SIGACCESS conference on computers and accessibility, Pittsburgh, PA, US

Felzer T, Strah B, Nordmann R (2008) Automatic and self-paced scanning for alternative text entry. In: Proceedings of the IASTED international conference on telehealth/assistive technologies, Baltimore, MD, US

Harbusch K, Kühn M (2003) An evaluation study of two-button scanning with ambiguous keyboards. In: Proceedings of AAATE'03

Koester HH, Levine SP (1994) Learning and performance of able-bodied individuals using scanning systems with and without word prediction. Assist Technol 6(1):42–53

Lin YL, Wu TF, Chen MC, Yeh YM, Wang HP (2008) Designing a scanning on-screen keyboard for people with severe motor disabilities. In: Miesenberger K, Klaus J, Zagler W, Karshmer A (eds) Computers helping people with special needs, LNCS, vol 5105. Springer, Heidelberg

Mackenzie IS, Felzer T (2010) SAK: scanning ambiguous keyboard for efficient one-key text entry. ACM Trans Comput-Hum Interact 17(3):11:1–11:39

Miro-Borras J, Bernabeu-Soler P (2008) E-everything for all: Text entry for people with severe motor disabilities. In: Proceedings of the 6th CollECTeR

Polacek O, Míkovec Z, Slavik P (2012) Predictive scanning keyboard operated by hissing. In: Proceedings of the AT'12, IASTED

Schadle I (2004) Sibyl: AAC system using NLP techniques. In: Miesenberger K, Klaus J, Zagler W, Burger D (eds) Computers helping people with special needs, LNCS, vol 3118. Springer, Heidelberg

Wandmacher T, Antoine JY, Poirier F, Départe JP (2008) Sibylle, an assistive communication system adapting to the context and its user. ACM Trans Access Comput 1(1): 6:1–6:30

Whiteside J, Archer N, Wixon D, Good M (1982) How do people really use text editors? ACM SIGOA Newslett 3(1–2):29–40

A Combinatory Approach to Assessing User Performance of Digital Interfaces

P. K. A. Wollner, P. M. Langdon and P. J. Clarkson

Abstract Digital devices are often restricted by the complexity of their user interface (UI) design. While accessibility guidelines exist that reduce the barriers to access information and communications technology (ICT), guidelines alone do not guarantee a fully inclusive design. In the past, iterative design processes using representative user groups to test prototypes were the standard methods for increasing the inclusivity of a given design, but cognitive modelling (the modelling of human behaviour, in this instance when interacting with a device) has recently become a feasible alternative to rigorous user testing (John and Suzuki 2009). Nonetheless, many models are limited to an output that communicates little more than the assumed time the modelled user would require to complete the task given a specific way of doing so (John 2011). This chapter introduces a novel approach that makes use of the overlay of user modelling output (timings) onto a graphical representation of an entire UI, thereby enabling the computation of new metrics that indicate the relative inclusiveness of individual screens of the UI.

1 Introduction

Digital devices are often restricted by the complexity of their user interface (UI) design. While accessibility guidelines exist that reduce the barriers to access information and communications technology (ICT), guidelines alone do not guarantee a fully inclusive design. In the past, iterative design processes using representative user groups to test prototypes were the standard methods for increasing the inclusivity of a given design, but cognitive modelling (the modelling of human behaviour, in this instance when interacting with a device) has

P. K. A. Wollner (✉) · P. M. Langdon · P. J. Clarkson
Engineering Design Centre, Department of Engineering, University of Cambridge, Cambridge, UK
e-mail: pkaw2@cam.ac.uk

P. M. Langdon et al. (eds.), *Inclusive Designing*, DOI: 10.1007/978-3-319-05095-9_4,
© Springer International Publishing Switzerland 2014

recently become a feasible alternative to rigorous user testing (John and Suzuki 2009). Nonetheless, many models are limited to an output that communicates little more than the assumed time the modelled user would require to complete the task given a specific way of doing so (John 2011).

In this chapter, we present a novel way of combining the output of user simulations into a unified mathematical representation, allowing for a more holistic interpretation of the simulation results. From a universal access perspective, this contributes to the field through improved accessibility of product designs which employ this novel approach. This is achieved through a more complete understanding of models that are based on specific impairments, as postulated by Langdon and Thimbleby (2010) and further explored by work such as Biswas et al. (2012). While we present this method to be utilised in combination with cognitive architectures (toolkits that combine theories of cognition to simulate human behaviour), it also may be presented as a novel way of interpreting studies involving real users and their performance data.

This chapter outlines (i) the structure and function of cognitive architectures, (ii) how these architectures can be employed to simulate user behaviour whilst interacting with touchscreen devices, (iii) the limitations of the output data (particularly from a universal access perspective), (iv) a proposed extension based on the application of graph theory on the output data and (v) how this mathematical model can be communicated to designers and developers in order to inform universally accessible designs.

2 Cognitive Architectures

Cognitive architectures are toolkits that combine theories of cognition to simulate human behaviour. The simulations are not purely a model of the behavioural output of the human mind but rather aim to replicate the structural properties of the modelled system. They do not physically replicate the components of the system—rather, virtual machines replicate the behaviour and knowledge of humans. By employing these frameworks to 'experience' the stimuli of a specific human–machine interaction scenario and in turn 'act' upon these stimuli, valuable user performance data can be generated by the model. This data is often represented as timings, based on a predefined set of actions which the architecture simulates, evaluates and outputs as a time value, indicating how long a real user would require to complete the same task.

Employing cognitive architectures in the design of interactive products and services allows the designer to gain insight on user behaviour without requiring the physical presence of users. Using a modelling approach as a replacement for user testing provides multiple benefits for both designers and—through the creation of improved products—for users.

Benefits for the designer include a process that supports quicker design iterations through the reduction of time-consuming participant recruitment and testing.

Furthermore, due to the ability to extend the model to factor in elements of the simulated users' prior knowledge, the resulting performance data can be tailored to the specific abilities of a subset of real users. This supports the design of interfaces targeted specifically at particular impairments, aiming to fulfil the overarching goal of more inclusive designs.

Benefits for the user may be defined in a similar way; the product better meets the cognitive requirements of a subsample of users based both on the cognitive impairments the users may have and the impact prior experience (Langdon et al. 2010) may have on their ability to interact with a new interface design. Benefits for the user, once the design is completed, include an optimised experience that is designed with the specific user type in mind.

While there are a number of cognitive architectures that are suitable for the automated performance evaluation of UI designs, we base much of the subsequent discussion on the assumed utilisation of ACT-R, one of the most extensive and evolved cognitive architectures (Anderson et al. 2004). This is based on the environment in which the related testing is performed (see Wollner et al. 2013) and represents the framework in which the model introduced in this chapter will be tested and deployed.

Originally, more advanced modelling tools such as ACT-R (which includes capabilities for simulated knowledge acquisition, i.e. the ability to learn and act upon learned facts) required extensive experience with the modelling language and the underlying assumptions in order to simulate cognitive processes accurately.

This implies that the entirety of the interaction process and environment are translated into a form that the architecture can interpret. Hence, this requires experts in these architectures to manually perform the translation and makes the potentially valuable data output inaccessible to most designers and/or developers.

Recently, cognitive models have become more accessible to designers with limited or no modelling experience (Councill et al. 2003); examples of this include Salvucci and Lee's (2003) ACT-Simple which uses a KLM-GOMS-based (John et al. 2004) descriptive language to automatically translate a specific UI design into a form that can be utilised as the basis of the simulation environment. Despite this simplification, there is still an inaccessibility in the assessment and communication of the output of cognitive models. Large amounts of segmented user data, mainly based on individual timings of actions within the UI, are presented to the designer without further analysis.

2.1 Screen Flow Network

The basis for this chapter is extending and combining the output of cognitive architectures. This is provided by establishing a graphic representation of the interaction paradigms and interaction routes available on digital interfaces. We present two types of elements included in this abstraction of UI progression: (i) elements that represent individual screens and (ii) connections that represent

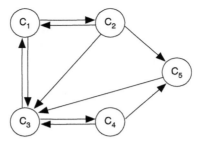

Fig. 1 In this screen flow, network circles represent individual screens within a UI progression, the lines represent the directionality of actions on individual screens that allow movement to another. E.g. C_1 has buttons or provides an implicit function to move to C_2 and C_3; C_5, in turn, only allows movement to C_3

actions that the user may complete when moving between these screens, including data input such as keyboard entry or drawing on the screen. The preliminary work (before running a cognitive model) builds this representation, the subsequent simulation extends the aforementioned connections between UI screens with values that represent the time a simulated user may require to move between screens. We refer to this as a 'screen flow network', as depicted in Fig. 1.

Despite the fact that the network displayed in Fig. 1 is not fully representative of a UI (in lacking the option to include global actions, such as a 'home' button and/or virtual global actions, such as an on-screen keyboard), it enables the representation of most actions that are based on the transition from one UI screen to the next. Further work will gradually extend the screen flow network to allow the representation of global actions.

Through the network introduced above, the designer can, through visual methods, explore design alternatives based on hot spots within the UI progression (indicated by timings). While this improves the design process, it cannot represent the entire network of possible screen flows the user may choose to explore. This is because the cognitive modelling approach is limited to a specific screen progression that the simulated user navigates through rather than making all screen flows possible. This means that the screen flow network provides a visual model indicating all possible screen flows given a specific UI design, but the modelled output of the same UI design is limited by the restricted information cognitive architectures can provide.

3 Simulation Network Analysis

Given the mismatch of the available and exploited data presented in the previous section, we propose to use a graph theoretical approach to make use of the overlay of user modelling output (timings) onto the entire screen flow network. This is

Fig. 2 Intersection network
used in factor graph notation,
where the arrows indicate
actions that allow the
navigation between screens
(denoted by C_n) and the
variables β_{n_1,n_2} are defined by
the simulation output

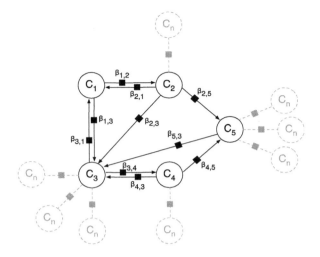

possible through the mapping of individual screens and their connections. We
propose that the user modelling is executed on all screen progressions (rather than
one specified route) and the resultant progression timings are included as weightings of the connecting elements of the screen flow network.

More specifically, we propose to integrate the availability of rich data regarding
the interconnectedness of individual screens and the sparse timings-based data that
is available as an output of user modelling of interfaces. Vertices in this representation match individual screens within the context of the application; edges are
the actions the user may complete to move between these screens. The edges are
reweighted by each iteration of implicit modelling (user simulations), allowing—
once a critical mass of simulations has been reached—a representation of both user
progression and potential (timing-based) performance bottlenecks within.

Figure 2 outlines the factor graph approach in more detail. The vertices C_n
represent individual screens within the UI flow. Screens are connected by edges
that have a weighting β_{n_1,n_2}, which represents the relative complexity of moving
from vertex C_{n_1} to C_{n_2}. The diagram indicates only one-directional weightings, in
cases, where there are no actions available to move to another screen. In the
example shown in Fig. 2, this applies to screen C_5, which can, for instance, only be
reached from screens C_2 and C_4. Hence the weightings $\beta_{2,5}$ and $\beta_{4,5}$ are non-zero.
The weightings $\beta_{5,2}$ and $\beta_{5,4}$ are zero and hence not depicted in the diagrammatic
representation. In contrast, screen C_5 only provides a direct action to move to
screen C_5, which, in turn, is indicated by the non-zero weighting $\beta_{5,3}$.

In this context, we need to define the factor graph theoretical relationships that
govern the weightings introduced in Fig. 2.

3.1 Longitudinal User Progression (β_{n_1,n_2})

The probabilistic interpretation that a user will experience difficulties when transitioning from screen C_{n_1} to screen C_{n_2} may be represented by the longitudinal user progression introduced in this section. By determining the timing of an individual screen change within the screen flow network, this descriptor can be numerically assessed based on the distribution of timings across all actions in the network. The descriptor is outlined in Eq. (1) where S is the set of all possible timed actions (edges) within the screen flow network and t_{n_1,n_2} is the timing for transitioning from vertex C_{n_1} to C_{n_2}.

$$\beta_{n_1,n_2} = \left[\frac{\sum_{i \in S} t_i}{|S|\left(t_{n_1,n_2}\right)} \right] \tag{1}$$

In other words, the descriptor introduced compares the simulated time an individual action takes in relation to the mean time of all available actions in that screen flow. This descriptor does not account for variations in user type.

3.2 Latitudinal User Progression ($\beta_{n_1,n_2}(u)$)

The probabilistic interpretation of the relative difficulty of a specific screen action for a specific user type u, based on the overall results is defined as latitudinal user progression. This is determined by comparing the variation in user timings based on the non-linear difference of timings of all user types, adjusted linearly by $\alpha(u)$, and compared to the mean timing of that action.

$$\beta_{n_1,n_2}(u) = \left[\frac{\sum_{v \in U} t_{n_1,n_2}(v)}{\alpha(u)|U|\left(t_{n_1,n_2}(u)\right)} \right] \tag{2}$$

In this representation, $\alpha(u)$ (defined in Eq. 3) is a function to correct for the specific user type, U is the set of all tested user types and $t_{n_1,n_2}(u)$ is the timing for transitioning from vertex C_{n_1} to C_{n_2} for user u.

$$\alpha(u) = \left[\frac{\sum_{v \in U} \sum_{i \in S} t_i(v)}{|U| \sum_{l \in S} t_l(u)} \right] \tag{3}$$

S is the set of all possible timed actions (edges) within the screen flow network and t_i is the timing for transitioning within vertex i.

In other words, the descriptor introduced in Eq. (2) is determined by the simulated time required for one action to be completed by one user type compared to the mean time of the same action for all other user types. This mean time is corrected to account for the overall (non-task specific) performance differences between user types by $\alpha(u)$ (defined in Eq. 3).

3.3 Comparison of Models

For both models, we have a simple distribution of values that allows the assessment of individual stages within the screen progression:

$\beta_{n_1,n_2} = 0$ It is impossible to progress from vertex C_{n_1} to C_{n_2}

$0 < \beta_{n_1,n_2} < 1$ It is difficult to transition from vertex C_{n_1} to C_{n_2}

$\beta_{n_1,n_2} = 1$ It is possible to transition from vertex C_{n_1} to C_{n_2}

$1 < \beta_{n_1,n_2}$ It is easy to transition from vertex C_{n_1} to C_{n_2}

Furthermore, there are subtle differences between the two models introduced in the sections above; while the first model is defined for only one user type, it is limited to compare performance timing only to all other actions within the same screen flow network.

The second model utilises a descriptor that compares the performance of a particular action within the screen flow network across all user types that were simulated.

While the first model highlights individual elements of the screen flow network that will—in relative terms—be more complex to a user and hence improves the usability across an interaction session, the second model assesses the complexity across user types and is more relevant for ensuring an inclusive design of the simulated interface.

Both models are constrained by utilising solely the mean timings (of all actions for one user type and for one action across all user types, respectively) rather than integrating the distribution of timings of these measures.

3.4 Complexity Propagation

Using the above-defined complexity models, we may define a formulaic approach for assigning complexity values to the vertices C_n based on the weightings (β) of the edges between them. For this, we resort to a message passing algorithm. If we assume the overall network presented in this chapter to be a bipartite graph where all nodes are either screens, denoted by the set C, or connections between screens (β), we may use the concept of belief propagation to evaluate the values of all screens. We call this concept complexity propagation. Here, to align the notation with that of Pearl (1982), we propose to denote screens as factors U and the movement between screens as variables V.

First, it is necessary to define a joint mass function, which is determined by the weighting of connections between screens, given a current valuation of the screen node they are connected to.

Next, the entire network is evaluated, by setting the values of each screen node, C_n, to a initialisation value and a message passing algorithm is employed until convergent values of the screen nodes are reached. This means that the screen nodes are revalued based on messages based through all neighbouring factor nodes β_{n_1,n_2}. These complexity messages are denoted as $\mu_{C \to \beta}$ and $\mu_{\beta \to C}$, respectively. The governing equations for this process are defined by Pearl (1982) (and subsequent work) and extend beyond the scope of this chapter.

4 Design Process Integration

Using the resultant graphical model permits the exploitation of various graph theoretical methods (Bondy and Murty 2008) that allow not only the definition of optimal routes of UI progression but also the topographical mapping of UI complexity, a method that provides a simplistic, data-rich representation of user complexity through screen flow networks. Three-dimensional mapping adds another dimension to the screen flow representation in edge and vertex form. The same concept as a standard network is utilised, but the third dimension is qualified by the distribution of weightings of all connected edges. Hence, the most altitudinal points on the visual representation relate to vertices (or screens) that are the most difficult to reach.

This representation provides a model for UI designers to better understand the limitations of a proposed design (given the simulation output thereof) without abstracting the key output of user modelling: timing-based data. While previous work (such as Thimbleby 2010) has suggested a purely graph theoretical approach to UI assessment, we propose a direct interpretation based on the timings-based output of cognitive models.

Here, we can take the final values of factor nodes (the values of screens C_n after the previously introduced complexity passing algorithm converges) to indicate the assumed complexity of a specific screen flow. This means that the relative complexity value established may be presented visually to the designer, allowing him or her to better comprehend the output of the simulation, translation and complexity passing procedure introduced in this chapter.

This could be visualised in a simple three-dimensional framework, where the x- and y-coordinates are determined arbitrarily (similarly to the screen flow network introduced earlier in this chapter) and the z-coordinate is established by a normalised representation of the factor value, post-convergence. A simplified example of such a representation is depicted in Fig. 3.

Fig. 3 Three-dimensional
complexity representation of
the screen flow network
introduced in Fig. 1, where
the height (along the *z-axis*)
defines the complexity of
reaching the given screen. In
this example, screen C_2 is the
one that is most difficult to
reach

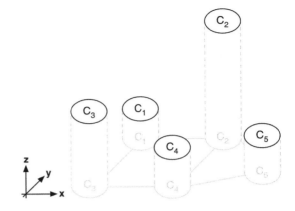

5 Conclusion

This chapter presents the mathematical foundation of a graph theoretical approach
for assessing UI complexity, given the timings-based output of cognitive simulations. It outlines (i) the framework in which a screen flow network is constructed,
(ii) the method by which individual cognitive simulations are instantiated and how
algorithmically all possible flows are processed, (iii) the approach by which the
resultant user progression data (i.e. timings) may be agglomerated into weightings,
(iv) the way in which probability of user access, β_{n_1,n_2}, may be identified and (v)
recommendations for further graph theoretical operations, given the completed
network.

We believe that this novel approach supports the design process of inclusive
user interfaces through two key mechanisms: representation and adaptability.
Representation is the process in which segmented numerical output data (i.e.
timings from cognitive architectures) is collated in a visual representation which
supports the designer in making better decisions based on data that gives insights
to a broad range of users. Adaptability relates to the underlying strength of cognitive architectures, which allow specific impairments to be modelled, giving the
designer the ability to investigate the usability of an interface for specific user
types.

Further work will focus on the extension of the complexity passing algorithm
introduced in this chapter and in developing a standardised three-dimensional
representation similar to Fig. 3 in order to communicate the output data of this
method to designers. Additionally, the computational framework may be extended
to be utilised not only with the output of cognitive architectures but also with user
performance data gained by actual user tests. This would require an adaption of the
user type variable introduced earlier in this chapter, but would greatly extend the
applicability of user testing in the design process.

References

Anderson JR, Bothell D, Byrne MD, Douglass S, Lebiere C et al (2004) An integrated theory of the mind. Psychol Rev 111:1036–1060

Biswas P, Robinson P, Langdon PM (2012) Designing inclusive interfaces through user modeling and simulation. Int J Human-Comput Interact 28(1):1–33

Bondy A, Murty U (2008) Graph theory. In: Graduate texts in mathematics. Springer, New York

Councill IG, Haynes SR, Ritter FE (2003) Explaining soar: analysis of existing tools and user information requirements. In: Proceedings of the 5th international conference on cognitive modeling

John BE (2011) Using predictive human performance models to inspire and support UI design recommendations. In: Proceedings of the ACM CHI conference on human factors in computing systems, Vancouver, Canada

John BE, Prevas K, Salvucci DD, Koedinger K (2004) Predictive human performance modeling made easy. In: Proceedings of the SIGCHI conference on human factors in computing systems, New York, NY, US

John BE, Suzuki S (2009) Toward cognitive modeling for predicting usability. In: Human-computer interaction. New trends. Lecture notes in computer science, vol 5610, pp 267–276

Langdon PM, Persad U, Clarkson PJ (2010) Developing a model of cognitive interaction for analytical inclusive design evaluation. Interact Comput 22(6):510–529

Langdon PM, Thimbleby H (2010) Inclusion and interaction: designing interaction for inclusive populations. Interact Comput 22(6):439–448

Pearl J (1982) Reverend bayes on inference engines: a distributed hierarchical approach. In: Proceedings of the American association of artificial intelligence national conference on AI, Pittsburgh, PA, US

Salvucci DD, Lee FJ (2003) Simple cognitive modeling in a complex cognitive architecture. In: Proceedings of the ACM CHI 2003 human factors in computing systems conference, Ft Lauderdale, FL, US

Thimbleby H (2010) Press on - Principles of interaction programming. MIT Press, Cambridge

Wollner PKA, Hosking I, Langdon PM, Clarkson PJ (2013) Improvements in interface design through implicit modeling. In: Universal access in human-computer interaction. Design methods, tools, and interaction techniques for eInclusion. Lecture notes in computer science,vol 8009, pp 127–136

How Interface Adaptation for Physical Impairment Can Help Able Bodied Users in Situational Impairment

P. Biswas, P. M. Langdon, J. Umadikar, S. Kittusami and S. Prashant

1 Introduction

Systems and services developed for elderly or disabled people often find useful applications for their able-bodied counterparts—a few examples are mobile amplification control, which was originally developed for people with hearing problems but is helpful in noisy environments, audio cassette versions of books originally developed for blind people, standards of subtitling in television for deaf users and so on. In this study, we evaluate how prediction from a user model developed for physically impaired users can work in situational impairment. This study aims to compare different interface layouts for touch screen hand-held devices and proposes modifications to those interfaces to make them more usable. Human machine interaction is different in hand-held devices from desktop or laptop computers as people use them while walking. Kane et al. (2008) proposed automatic interface zoom when users interact with the device while walking and found bigger target and font sizes tend to facilitate interaction. Schildbach and Rukzio (2010) took those results forward and conducted an ISO 9241 pointing task while users were walking. They also proposed bigger target sizes but at the same time noted that bigger font size did not reduce reading time due to time spent in scrolling. Macik et al. (2013) proposed a more formal architecture for context-based adaptation than zooming alone which also addressed the needs of users with physical impairment. The SUPPLE project (2009) personalises interfaces mainly by changing layout and font size for people with visual and motor impairment and also for general use. However, the user models do not consider visual and motor

P. Biswas (✉) · P. M. Langdon
Engineering Design Centre, Department of Engineering, University of Cambridge, Cambridge, UK
e-mail: pb400@eng.cam.ac.uk

J. Umadikar · S. Kittusami · S. Prashant
IITM's Rural Technology and Business Incubator, Chennai, India

P. M. Langdon et al. (eds.), *Inclusive Designing*, DOI: 10.1007/978-3-319-05095-9_5,
© Springer International Publishing Switzerland 2014

impairment in detail and thus work only for loss of visual acuity and a few types of motor impairment.

This study takes an inclusive approach—it uses an inclusive user model (Biswas and Langdon 2012; Biswas et al. 2012), which has a more sophisticated mechanism for mapping users' range of abilities to interface elements than Macik's (2010) framework and considered more types of users than SUPPLE (The SUPPLE project 2009). In this work, it has been used to adapt interfaces to address situational impairment. We have developed a set of models that can predict human machine interaction for different types of visual, hearing and motor impairment (Biswas and Langdon 2012; Biswas et al. 2012). Previous studies have already found the effect of bigger target sizes, so this study kept buttons of similar sizes in both adapted and non-adapted conditions. This study also took a more generalised approach than Kane et al. (2008) and Schildbach and Rukzio (2010). Kane worked on a music player interface while Schildbach conducted an ISO 9241 task. This study investigated different mobile applications and identified two different screen layouts. We developed prototypes of adapted and non-adapted versions of those screen layouts and investigated target selection times and numbers of wrong selections in different conditions including users with colour blindness. In the adapted version, we changed font size, colour contrast and button spacing keeping the target sizes unchanged.

2 Background

The study is conducted in the context of the IUATC project. The India-UK Advanced Technology Consortium (IUATC 2013) is developing a number of applications to improve the quality of life of the rural population. We have considered two applications here which will be deployed on ubiquitous or hand-held devices.

The pest-disease image upload (PDIU) application will be used by farmers to upload images of infested crops, while they are in the field. The uploaded images will automatically be sent to remotely located experts, who will advise farmers about the required treatment. The application not only aims to make it easy for farmers who have difficulty in operating the keypad but also accommodates those suffering from poor vision or cognitive impairments.

The chatterbox application is part of an electronic health-management system and aims to develop a social media like system for rural elderly users. It can be used by either patients or caregivers or any other group of people to interact among themselves, access information and use relevant health services.

3 Participants

We collected data from nine users (five male, four female) with an average age of 28.78 years. They all volunteered for the study and none had any difficulty in undertaking the experimental tasks.

4 Design

We conducted a visual search and touch task involving a tablet. The screens were designed to be similar to the applications developed in the IUATC project though they were quite similar to standard Windows interfaces. There were two different arrangements of buttons—one arrangement was like a home screen with square buttons two in a row (tile layout, Fig. 1), the other arrangement was like a list of menu items or textboxes (List layout, Fig. 2). We also had adapted versions of these two screen layouts. For each type of screen layout each participant undertook eight search and click tasks.

The participants undertook the trial while walking in a field (Fig. 3). The field does not have any traffic or other obstruction that can physically hurt the participant. All trials took place in daylight and in the same place. Participants did not report any issue with the legibility of the screen.

The non-adapted version had default font size (8.25 point) and the colour contrast used in Microsoft Visual Studio 2010 and no spacing between screen elements (Figs. 4 and 5). The adapted version has 12 point font size and 6 pixel padding between any two screen elements (Figs. 4 and 5). The button spacing was chosen to keep a visible separation among screen elements as well as keeping a minimum separation predicted by our simulator to avoid missed clicks. The colour contrast was adjusted according to the type and presence of colour blindness of participants—Blue–Yellow for Red–Green colour blindness and Black–White for any other type of colour blindness. The first three experimental screens were 200 × 400 pixels making them roughly equivalent to a 1.3 in. × 2.7 in. screen area, which is approximately the same size as the screen area devoted to the menu items in a 3.7 in. display (like Nokia Lumia 800). The adapted version of the list screen layout was 465 pixels long to accommodate all screen items.

Figure 6 demonstrates the effect of moderate visual impairment, i.e. loss of acuity and red–green colour blindness on the perception of interfaces. It can be seen that the adapted version becomes legible for these moderate visual impairments. A similar analysis was conducted for motor impairment as well. The study evaluates how adaptations for physical impairment can be beneficial for situational impairments.

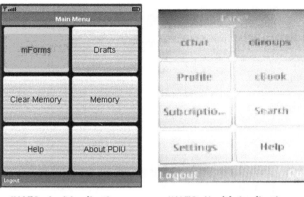

IUATC eAgri Application   IUATC eHealth Application   Nokia Lumia Interface

Fig. 1 Existing tiled interfaces

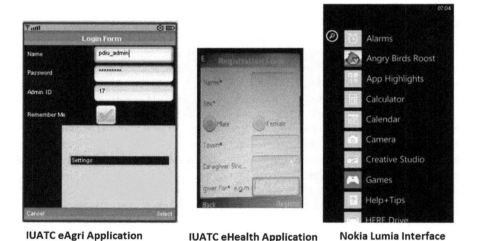

IUATC eAgri Application   IUATC eHealth Application   Nokia Lumia Interface

Fig. 2 Existing list interface

5 Material

The experiment was conducted on Acer Iconia tablets with 10 in. screen
800 × 1280 pixels of resolution and running Windows 7 operating system. Participants used their fingers to make a selection on the touch screen holding the
tablet in vertical position. We locked the screen to avoid any unintended rotation.

Fig. 3 Experimental set up

Fig. 4 Experimental tiled interface

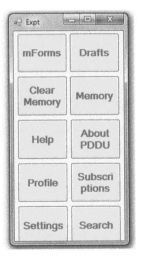

Fig. 5 Experimental list interface

Non Adapted List Screen

Adapted Version (for Red-Green Colour-Blindness)

Fig. 6 Visual impairment simulation on tiled interface; similar to Fig. 4

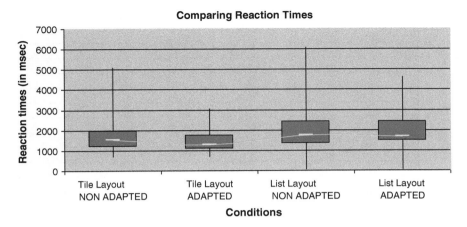

Fig. 7 Comparing selection times

6 Procedure

Initially we tested participants for colour blindness using plates 16 and 17 of Ishihara Colour Blindness Test. We also checked their visual acuity to confirm they could read the screen. Next participants were briefed about the task and once they were satisfied they undertook the trial. We developed a software that showed a button with a caption to participants which once pressed showed the experimental screen. We recorded the reaction time between onset of the experimental screen and the selection of the button. The button captions were taken from the IUATC applications and the order of different experimental conditions was randomised across participants.

7 Results

We found that the selection time was lowest which the adapted tiled layout (Fig. 7). A Layout × Adaptation (2 × 2) ANOVA on selection times found a significant effect ($F(1,248) = 36.08$, $p < 0.01$) of the type of layout, i.e. the tiled layout is better than the list layout. Missed clicks were highest for the non-adapted list layout while there was no wrong selection, for the adapted tiled layout (Fig. 8). An unequal variance t-test found that users took significantly ($p < 0.05$) less time to select on the adapted tiled version than the non-adapted tiled version though the selection times are not significantly different between adapted and non-adapted versions of the list layout. However, the number of wrong selections was lower with the adapted list layout than the non-adapted version.

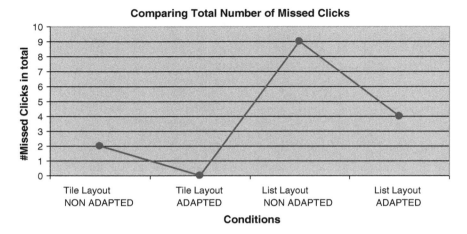

Fig. 8 Comparing number of wrong selections

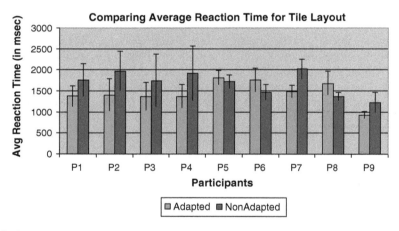

Fig. 9 Comparing average individual selection time for tiled layout

We also conducted a person-by-person analysis and found that users took less time on average to select the correct button in the adapted tiled version than the non-adapted one (Fig. 9). Among nine users, six took less time on average to select correct target. For list layout, the distribution was rather even, with four users taking more time in the adapted version, four vice versa and one user taking almost equal time in both (Fig. 10).

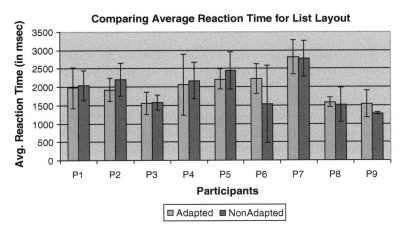

Fig. 10 Comparing average individual selection time for list layout

8 Discussion

This study aims to evaluate how an inclusive user model can be used to adapt interfaces even for able-bodied users with situational impairment. We considered two common types of touch screen interfaces and found that an adapted version either reduces selection times or the number of wrong selections. The study also reveals that the tiled layout is better than the list layout and a bigger font and padding between items reduces both chances of wrong selection and selection times.

The present study used only texts but conventional touch screen interfaces also have icons. However, the label matching principle (Rieman and Young 1996) says that visual search is often governed by textual labels for novice users. With the rapid growth of ubiquitous devices, users change their mobile handset pretty often and a huge portion of users can be considered novices with respect to a new device or interface.

In this study, we determined the button size by the length of textual captions. The buttons were big enough to accommodate the biggest label. The adapted list layout needs a bigger screen size to avoid the need for scrolling and to keep the selection times comparable to the other three conditions. Although different IUATC and other existing applications have used different button sizes, our results can be extrapolated to any size of button. If an interface contains very big buttons then the overall selection time may be reduced due to time devoted to scrolling, and too small a button size will limit the font size available to accommodate users with either visual or situational impairment. However, insertion of padding between buttons or clickable screen elements may not consume too much screen space and was found to reduce the number of wrong selections.

9 Conclusions

We have developed a user model to adapt interfaces for users with physical impairment. In this paper, we report results from a user study about the performance of the user model while adapting interfaces for situational impairment. In particular, we conducted a study of visual search and selection tasks with a touch screen tablet while users were walking. We found that an adapted interface with bigger font size and button spacing can reduce selection time and chances of wrong selection for existing touch screen interface layouts.

References

Biswas P, Langdon PM (2012) Developing multimodal adaptation algorithm for mobility impaired users by evaluating their hand strength. Int J Hum Comput Interact 28(9):576–596

Biswas P, Langdon PM, Robinson P (2012) Designing inclusive interfaces through user modelling and simulation. Int J Hum Comput Interact 28(1):1–33

IUATC (2013) The IUATC consortium. www.iu-atc.com. Accessed 29 Oct 2013

Kane SK, Wobbrock JO, Smith IE (2008) Getting off the treadmill: evaluating walking user interfaces for mobile devices in public spaces. In: Proceedings of the 10th conference on human-computer interaction with mobile devices and services, Amsterdam, The Netherlands

Macik M, Cerny T, Basek J, Slavik P (2013) Platform-aware rich-form generation for adaptive systems through code-inspection. In: Human factors in computing and informatics, Lecture Notes in Computer Science, vol 7946. Springer, Berlin

Rieman J, Young RM (1996) A dual-space model of iteratively deepening exploratory learning. Int J Hum Comput Stud 44:743–775

Schildbach B, Rukzio E (2010) Investigating selection and reading performance on a mobile phone while walking. In: Proceedings of the 12th conference on human-computer interaction with mobile devices and services, Lisbon, Portugal

The SUPPLE Project (2009) SUPPLE: Automatic generation of personalizable user interfaces. www.cs.washington.edu/ai/supple/. Accessed 29 Oct 2013

Gender Issues in ICT Adoption: A Literature Review

T. S. Goldhaber, P. M. Langdon and P. J. Clarkson

Abstract In the UK, only 66% of people over the age of 65 and 29% of people over the age of 75 have used the Internet. While it is important to design inclusive Information and Communication Technology (ICT) for this population, "older people" are often lumped together by designers, negating the diversity within this population. One important factor that is not taken into account is gender differences during learning. There is a rich history of research in how men and women react differently to various aspects of the learning experience, particularly success and failure, but this knowledge has yet to be incorporated into many ICT designs. This paper discusses the gender differences literature as it relates to user interface design for the aging population.

1 Introduction

As more and more critical activities and services move online, digital inclusion has become a societal imperative. One highly excluded group remains the "ageing population", an ill-defined set of people ranging in age from about 60 to well over 90 from every imaginable socio-economic and cultural background. Broadly, it is fair to say that older people tend to be excluded by modern technology: 44 % of people over the age of 65 and 71 % of people over the age of 75 in the UK have not used the internet (Ofcom 2011; Office for National Statistics 2012; Lewis 2013). However, it is important to keep in mind that the reasons behind lower adoption rates are not uniform, and that inclusive design as a field must strive to understand the diversity of excluding factors.

T. S. Goldhaber (✉) · P. M. Langdon · P. J. Clarkson
Engineering Design Centre, Department of Engineering, University of Cambridge, Cambridge, UK
e-mail: tg313@cam.ac.uk

P. M. Langdon et al. (eds.), *Inclusive Designing*, DOI: 10.1007/978-3-319-05095-9_6,   59
© Springer International Publishing Switzerland 2014

One factor that often is not taken into consideration is gender. Studies have uncovered gender differences in technology learning and motivation (e.g. Harter 1978; Malone 1981; Beckwith and Burnett 2004; Burnett et al. 2008; Goldhaber et al. 2013), finding, for example, that women tend to be motivated by feelings of competence or enjoyment of an activity whereas men can be motivated by feeling challenged or frustrated. However, there has been a paucity of design research on the underlying causes of these differences. Although not often referenced in the design literature, there is a rich history of gender research with a particular focus on learning and learning differences. This paper reviews a selection of gender research particularly relevant to design of ICT for the ageing population.

2 Gender Differences in Self-Efficacy

Self-efficacy is the belief a person has that he or she can successfully carry out a task. For learning, it is an incredibly important construct because feelings of capability are related to how well someone learns and what tasks they pursue. For example, Postkammer and Smith (1985) found that self-efficacy was 'a major predictor of interests' for junior high school students (with gender differences already starting to appear at that age). Betz and Hackett also have a series of studies over a decade demonstrating that lower self-efficacy in mathematics reduces the likelihood that an individual will pursue a maths-related field in further study (Betz and Hackett 1981, 1983; Hackett and Betz 1989).

This is not just related to mathematics, however, but to the application of effort more generally. Boggiano et al. (1988) showed that children's sense of competence and autonomy in an activity facilitated their preference for increased challenge. Therefore, increasing self-efficacy during learning should be a focal point of ICT design efforts.

There is a plethora of literature demonstrating that women tend to have lower self-efficacy and achievement expectancies compared to men and rate their performance as worse than men even when their abilities are equal (Crandall 1969; Betz and Hackett 1981; Gitelson et al. 1982; Erkut 1983; Miura 1987; Sleeper and Nigro 1987; Beyer 1990). In particular, there appears to be a much smaller correlation between self-efficacy and ability for women as compared to men (Betz and Hackett 1981; Miura 1987), and in general women will underestimate their ability whereas men will overestimate theirs (Beyer 1990). This can also be a hard problem to rectify for women, with one study finding that a 'calibration' task (a pre-test on which participants were given feedback) had an effect on men but not on women (Dargnies and Hollard 2009). Women also tend to make external attributions for success, whereas men credit their ability (Erkut 1983; Beyer 1990).

The differences seem to be more severe in certain situations. For example, women have lower self-efficacy for tasks or occupations seen as traditionally male or masculine (Betz and Hackett 1981, 1983; Hackett and Betz 1981, 1989; Lent et al. 1984; Hackett 1985; Postkammer and Smith 1985; Busch 1995, 1996) such as mathematics and engineering (Betz and Hackett 1981) and computer skills (Miura 1987; Vasil et al. 1987; Jorde-Bloom 1988; Harrison and Rainer 1992; Busch 1995, 1996). Moreover, while women had more or less equivalent self-efficacy to men for simple computer tasks, their self-efficacy was much lower than men for complex computer tasks (Murphy et al. 1989; Busch 1995), for example, calibrated simple versus complex tasks in a word processor in Busch (1995). This historically has contributed to lower numbers of women in computing fields (e.g. Klawe and Leveson 1995; Gürer and Camp 2002; Klawe et al. 2009), which of course has not helped the view of computing as a masculine field.

In summary, there is a rich body of literature that demonstrates deep gender divides in self-efficacy. Because self-efficacy is so important for learning, however, it is important to understand and attempt to rectify self-efficacy differences in a design context.

3 Differences in Competitiveness, Risk Aversion and Affinity for Challenge

Probably because of women's lower self-efficacy, they also tend to be less competitive, more risk averse, and more averse to high levels of challenge (Byrnes et al. 1999; Niederle and Yestrumskas 2008; Booth 2009). Niederle and Yestrumskas (2008) found that, given a choice between an easy or hard task, women sought out challenge less than men. Related to this is women's increased risk aversion. Byrnes et al. (1999) found that men take too many risks, whereas women decline to take risks even in safe, consequence-free environments and situations. This may also affect how men and women respond to different kinds of rewards and feedback (Datta Gupta et al. 2005). Finally, women are less competitive than men (Niederle and Vesterlund 2007), with men more likely to compete against other competitive people in a tournament situation (Datta Gupta et al. 2005) and performing better in competitive situations (Gneezy et al. 2003).

While women tend to shy away from competition, they also have a more averse reaction to challenge or confusion. For example, in a group of high-achieving men and women of equal ability, men improve their effort and performance after a period of confusion or challenge, whereas women's performance gets worse (Licht and Dweck 1984; Dweck 2006). This is very relevant to learning new technological skills, because these results indicate that women are particularly susceptible to failure early on during learning, a situation that has yet to be rectified in the majority of ICT designs.

4 Achievement Orientation

Why are women turned off by failure, confusion, and challenge while men are motivated by it? It might be a question of how ability and learning is viewed by an individual.

Dweck (2006) writes about the tendency of girls to view ability as innate and unchangeable, whereas boys view ability as flexible and as changing with increased effort. A rigid view of ability is detrimental to overcoming challenges or pursuing an activity past an initial failure since an innate and unchangeable ability is unlikely to respond to increased efforts. Someone with a rigid self-image of their ability is therefore likely to see failure as an indication that they are not suited to the task at hand, not that they should try harder.

Contrary to popular educational beliefs, it appears that encouraging students based on intelligence or talent actually worsens the problem. Says Dweck: *"In a series of studies, we have shown that praising students' ability (even after a job truly well done) has a host of undesirable consequences* (Mueller and Dweck 1998). *First, it conveys to them that their ability is a gift and makes them reluctant to take on challenging tasks that hold a risk of mistakes. Next, when these same students hit a period of difficulty, the ones who had been praised for their ability tended to lose their confidence. If their success meant they had the gift, their struggles now told them they didn't. As a result, they lose interest in pursuing the task (just like females and math) and show a sharp decline in their performance* (Dweck 2006)."

However, it is possible to encourage a flexible self-view. Hong et al. (1999) and Blackwell et al. (2007), for example, both showed that teaching about flexible ability and downplaying the effects of innate 'gifts' improved achievement over the long-term in students. Nonetheless, older people today—and particularly women—who are trying to learn to use computers did not have the benefits of this insight in their earlier years and must be doubly encouraged.

Goldhaber et al. (2013) conducted a study in which people over the age of 60 were asked to play either an easy but boring game, an easy and interesting game, or a very challenging but interesting game. The results uncovered many gender differences, but in particular that when given a choice to keep playing the game after the official end of the game play period, women continued to play the game that they had enjoyed and had made them feel competent, whereas men only continued to play the very challenging game. A rigid ability view for women and a flexible one for men helps to explain why men would spend the most time playing a game on which they did not do well and which they did not enjoy. Men chose to redouble their efforts in response to failure whereas women spent more time playing the games at which they felt most competent. Dweck (2006) states that: *"We had found in our past research that viewing intellectual ability as a gift (a fixed entity) led students to question that ability and lose motivation when they encountered setbacks. In contrast, viewing intellectual ability as a quality that could be developed led them to seek active and effective remedies in the face of difficulty.*" The results from Goldhaber's et al. (2013) study appear to be largely in line with Dweck's findings.

The work of Dweck and others indicates that if there were a way to encourage a more flexible self-view in users, it could have huge implications for ICT learning. What is clear is that design must take into consideration the need to emphasise flexible learning and rewarding effort. An interface with low levels of transparency or feedback in particular has the potential to de-motivate individuals with a rigid achievement orientation.

5 Overconfidence, Underperformance and Ego Utility

Koszegi (2006) introduces the concept of 'ego utility'—the importance of positive self-image to an individual and how that motivates their actions. *"Overconfidence in beliefs can be coupled with underconfidence in observed actions. In fact, among people who are identical other than in the importance they attach to ego utility, there is a negative relationship between confidence and participation in the ambitious task: Although average overconfidence increases in the weight attached to ego utility, the probability of choosing the ambitious option approaches zero as that weight approaches infinity,"* Koszegi explains. In other words, an individual might not choose a task that he or she feels confident about precisely because he or she does not want to be proven wrong or shown that the task is in fact beyond his or her ability. Similarly but counter-intuitively, someone with high ego utility might continue with a very hard task (at which they are not doing well) to attempt to prove their ability in the hopes of maintaining their ego.

An alternate way of explaining this is that an individual will keep 'collecting data' about themselves by doing an activity only while the data does not reinforce their own self-perception. Someone with high self-efficacy will see a positive result as a reinforcement of their high ability and stop collecting data, whereas a negative result will necessitate a further effort until their achievement becomes consistent with their perceived ability level. Conversely, someone with low self-efficacy will see a negative result as reinforcing their low ability level and stop collecting data, but unexpected positive results will motivate a further effort.

Koszegi's theory is supported by results from Rawsthorne and Elliot (1999) who found that study participants would voluntarily keep working on a task after negative feedback to 'demonstrate to themselves that they have the ability to perform well'. In addition, as men tend to be overconfident (Beckwith and Burnett 2004), this theory explains why women are often more likely to give up sooner when attempting a challenging task such as learning to use a new ICT device.

6 Culture Differences and Shifting Norms

It is worth noting that the majority of the studies on gender differences in learning and motivation have taken place in the West, so it is important to consider the role that culture has to play in determining the differences between men and women around the world. While the anthropological literature on this subject is largely beyond the scope of the present research, there are a few representative papers of note.

Datta Gupta et al. (2005) and Busch (1995) both note that prior experience and social norms likely have a huge impact on computer learning, with Datta Gupta emphasising the probable effect of social norms on women's competitiveness and Busch finding that women's more negative attitude towards computers was largely mediated by computer experience. However, there is increasing evidence of the cultural side of competitiveness. For example, Gneezy et al. (2009) compared gendered preference for competition in a patriarchal tribe in Tanzania and a matrilineal tribe in India. They found that women in the matrilineal tribe were far more competitive than the men, with levels of competitiveness equal or exceeding those of the men in the patriarchal tribe. Furthermore, Khatoon and Mahmood (2011) found that girls in India have higher computer self-efficacy than boys. Therefore, it is not unreasonable to assume that as gender differences become less pronounced in society, that the gender differences currently observed in the ageing population will decrease as well, but the issues present are nonetheless likely to be relevant for at least several more decades.

7 Conclusions and Design Implications

This work has several implications for inclusive ICT design for the ageing population. First and foremost, the clear gender differences that are present in reaction to challenge and failure necessitate strong design consideration. It is crucial that designers look at gender as a factor in collaborative design and user testing, for example. While it is certainly possible that there are designs that appeal or do not appeal to both genders, a universal appeal or suitability cannot be assumed. Therefore, the possibility of designing separately for men and women (or at least designing different learning experiences) should be considered. While it may not always be possible or desirable, if gendered design increases adoption and improves learning, this is something that designers very much need to take into account.

Second, it is clear that the early learning experience for new ICT needs to be carefully designed, particularly for women. Designing a first experience to be a simpler, easier version of the main task could decrease negative emotions associated with failure in women and prime them to see themselves as capable of doing the task. These sorts of interventions may not work as well for men, but if they increase learning and skill (e.g. Carroll and Carrithers 1984), they may still be valuable. In addition, in terms of user testing, it is critical to look at the time-course of learning, not just overall failure rate. In other words, failures or successes that occur within the first few minutes of a trial might have far more motivational impact than failures after ten minutes, and this must be taken into consideration when looking at overall usability and intrinsic motivation.

While gender differences in learning are likely to decrease in the coming decades, taking gender into account for those who are currently new to ICT has the potential to drastically increase adoption. Moreover, as women tend to live longer than men, enabling higher ICT use rates among women will further extend the benefits to the population as a whole.

References

Beckwith L, Burnett M (2004) Gender: an important factor in end-user programming environments? In: Proceedings of the visual languages and human centric computing, 2004 IEEE symposium, CA, US

Betz NE, Hackett G (1981) The relationship of career-related self-efficacy expectations to perceived career options in college-women and men. J Couns Psychol 28(5):399–410

Betz NE, Hackett G (1983) The relationship of mathematics self-efficacy expectations to the selection of science-based college majors. J Vocat Behav 23(3):329–345

Beyer S (1990) Gender differences in the accuracy of self-evaluations of performance. J Pers Soc Psychol 59(5):960–970

Blackwell LS, Trzesniewski KH, Dweck CS (2007) Implicit theories of intelligence predict achievement across an adolescent transition: a longitudinal study and an intervention. Child Dev 78(1):246–263

Boggiano AK, Main DS, Katz PA (1988) Childrens preference for challenge: the role of perceived competence and control. J Pers Soc Psychol 54(1):134–141

Booth AL (2009) Gender and competition. IZA Discussion Papers, No. 4300: 1–30

Burnett M, Wiedenbeck S, Grigoreanu V, Subrahmaniyan N, Beckwith L et al (2008) Gender in end-user software engineering. In: Proceedings of the 4th international workshop on End-user software engineering, Leipzig, Austria

Busch T (1995) Gender differences in self-efficacy and attitudes toward computers. J Educ Comput Res 12(2):147–158

Busch T (1996) Gender, group composition, cooperation, and self-efficacy in computer studies. J Educ Comput Res 15(2):125–135

Byrnes JP, Miller DC, Schafer WD (1999) Gender differences in risk taking: a meta-analysis. Psychol Bull 125(3):367–383

Carroll JM, Carrithers C (1984) Training wheels in a user interface. Commun ACM 27(8):800–806

Crandall VC (1969) Sex differences in expectancy of intellectual and academic reinforcement. In: Smith CP (ed) Achievement-related motives in children. Russell Sage Foundation, NY

Dargnies MP, Hollard G (2009) Incentives to learn calibration: a gender dependent impact. Econ Bull 29(3):1820–1828

Datta Gupta N, Poulsen A, Villeval MC (2005) Male and female competitive behaviour: experimental evidence. IZA Discussion Papers, No. 1833

Dweck CS (2006) Is math a gift? Beliefs that put females at risk. In: Ceci SJ, Williams W (eds) Why aren't more women in science? top researchers debate the evidence. American Psychological Association, Washington, DC

Erkut S (1983) Exploring sex-differences in expectancy, attribution, and academic-achievement. Sex Roles 9(2):217–231

Gitelson IB, Petersen AC, Tobinrichards MH (1982) Adolescents expectancies of success, self-evaluations, and attributions about performance on spatial and verbal tasks. Sex Roles 8(4):411–419

Gneezy U, Leonard KL, List JA (2009) Gender differences in competition: evidence from a matrilineal and a patriarchal society. Econometrica 77(5):1637–1664

Gneezy U, Niederle M, Rustichini A (2003) Performance in competitive environments: gender differences. Quart J Econ 118(3):1049–1074

Goldhaber TS, Langdon PM, Clarkson PJ (2013) Designing intrinsically motivating user interfaces for the ageing population. In: Proceedings of the 15th international conference on human-computer interaction (HCII), Las Vegas, NV, US

Gürer D, Camp T (2002) An ACM-W literature review on women in computing. ACM SIGCSE Bull Women Comput 34(2):121–127

Hackett G (1985) Role of mathematics self-efficacy in the choice of math-related majors of college-women and men: a path-analysis. J Couns Psychol 32(1):47–56

Hackett G, Betz NE (1981) A self-efficacy approach to the career-development of women. J Vocat Behav 18(3):326–339

Hackett G, Betz NE (1989) An exploration of the mathematics self-efficacy/mathematics performance correspondence. J Res Math Educ 20(3):261–273

Harrison AW, Rainer RKJ (1992) The influence of individual differences on skill in end-user computing. J Manage Inf Syst 9(1):93–111

Harter S (1978) Effectance motivation reconsidered toward a developmental model. Hum Dev 21(1):34–64

Hong YY, Chiu CY, Dweck CS, Lin DMS, Wan W (1999) Implicit theories, attributions, and coping: a meaning system approach. J Pers Soc Psychol 77(3):588–599

Jorde-Bloom P (1988) Self-efficacy expectations as a predictor of computer use: a look at early childhood administrators. Comput Schools, 5(1–2):45–63

Khatoon T, Mahmood S (2011) Computer attitude as a function of gender, type of school, mathematics anxiety and mathematics achievement. Eur J Soc Sci 18(3):434–443

Klawe M, Leveson N (1995) Women in computing: where are we now? Commun ACM 38(1):29–35

Klawe M, Whitney T, Simard C (2009) Women in computing—take 2. Commun ACM 52(2):68–76

Koszegi B (2006) Ego utility, overconfidence, and task choice. J Eur Econ Assoc 4(4):673–707

Lent RW, Brown SD, Larkin KC (1984) Relation of self-efficacy to career choice and academic performance. J Couns Psychol 31:356–362

Lewis G (2013) Digital inclusion policy review. AgeUK, London

Licht BG, Dweck CS (1984) Determinants of academic-achievement—the interaction of childrens achievement orientations with skill area. Dev Psychol 20(4):628–636

Malone TW (1981) Toward a theory of intrinsically motivating instruction. Cogn Sci 5(4):333–369

Miura IT (1987) The relationship of computer self-efficacy expectations to computer interest and course enrollment in college. Sex Roles 16(5–6):303–311

Mueller CM, Dweck CS (1998) Praise for intelligence can undermine children's motivation and performance. J Pers Soc Psychol 75(1):33–52

Murphy CA, Coover D, Owen SV (1989) Development and validation of the computer self-efficacy scale. Educ Psychol Measur 49(4):893–899

Niederle M, Vesterlund L (2007) Do women shy away from competition? do men compete too much? Quart J Econ 122(3):1067–1101

Niederle M, Yestrumskas AH (2008) Gender differences in seeking challenges: the role of institutions (Mimeo). Stanford University, Stanford

Ofcom (2011) Communications market report: UK

Office for National Statistics (2012) Internet access: Households and individuals

Postkammer P, Smith PL (1985) Sex-differences in career self-efficacy, consideration, and interests of 8th and 9th graders. J Couns Psychol 32(4):551–559

Rawsthorne LJ, Elliot AJ (1999) Achievement goals and intrinsic motivation: A meta-analytic review. Pers Soc Psychol Rev Off J Soc Pers Soc Psychol Inc 3(4):326–344

Sleeper LA, Nigro GN (1987) It's not who you are but who you're with—self-confidence in achievement settings. Sex Roles 16(1–2):57–69

Vasil L, Hesketh B, Podd J (1987) Sex-differences in computing behavior among secondary-school pupils. NZ J Educ Stud 22(2):201–214

Blind and Deaf Consumer Preferences for Android and iOS Smartphones

J. Morris and J. Mueller

1 Introduction

Access to and use of mobile wireless technology has become critical to social and economic participation for people with disabilities. As the technology increases in power and sophistication, these customers increasingly rely on mobile devices and software for functions previously available only through dedicated 'assistive technology'. Successfully serving this large and growing population has become a market imperative as well as a legislative mandate for the wireless industry in the US. Competition for this market is especially keen between the Android and Apple's iOS operating systems. This article presents survey research findings on the relative ease of use, importance and satisfaction experienced by blind and deaf customers using mobile devices on Android or iOS platforms. Focus group research conducted by the authors suggests that blind smartphone users overwhelmingly favour the iPhone, while deaf smartphone users show greater diversity in their device choices. Data collected through the Survey of User Needs (SUN) conducted by the Rehabilitation Engineering Research Center for Wireless Technologies (Wireless RERC) are presented to test this finding with quantitative data, and to test differences between blind and deaf users in terms of ease of use and satisfaction with iPhone and Android smartphones.

2 The Wireless Marketplace Meets Customers with Disabilities

In the US, accessibility and usability of wireless information and communication technologies (ICT) has been a legislative mandate for many years, most recently under the Twenty-First Century Communications and Video Accessibility Act

J. Morris · J. Mueller (✉)
Rehabilitation Engineering Research Center for Wireless Technologies, Atlanta, GA, USA
e-mail: jlminc1@verizon.net

P. M. Langdon et al. (eds.), *Inclusive Designing*, DOI: 10.1007/978-3-319-05095-9_7,   69
© Springer International Publishing Switzerland 2014

(CVAA), which began to take effect in 2013 (U.S. Government Printing Office 2010). Reinforcing this mandate, intense competition within segments of the ICT industry has made people with disabilities—including a large proportion of older citizens—an attractive customer market.

This competition is especially obvious between products and services offered on the Android and Apple wireless platforms, the two dominant smartphone and tablet operating systems. With each release of updated operating systems, debate is renewed about which platform serves customers better, with and without disabilities. Other design qualities, including the form factor (pure touchscreen vs. physical keyboard, 'candy-bar' vs. clamshell design, etc.), overall organisation of features and functions in operating system menus and out-of-the-box accessibility can greatly impact usability. The virtual controls of touchscreen mobile devices, and their varying placement, style, and functionality among diverse apps and operating systems, can present challenges to users with disabilities. Customers who need or prefer a tactile keyboard over a touchscreen can choose only between Android and Blackberry devices.

Long considered the gold standard of smartphone operating systems, Apple's iOS is being effectively challenged by new capabilities available on the Android operating system. In September, 2012, a Business Insider online column rated Google Now competitive with Siri, Apple's popular voice-activated user interface. The column also cited Android's Talkback, which provides audio feedback when navigating the touchscreen interface with your finger, as competitive with VoiceOver on iOS. On the other hand, the accessibility features of Apple's iOS 6, including Guided Access for users with cognitive disabilities, continues to out-shine those of Android Jelly Bean. (Smith 2012).

A 2013 comparison of Android 4.2 (Jelly Bean) and Apple iOS 6.1 operating systems (Ybanez 2013) noted the following:

The Android operating system offers this array of accessibility features:

1. TalkBack—provides voice feedback and navigating by swiping gesture with Explore by Touch feature;
2. magnification gestures—magnifies the screen with swiping gestures;
3. large text—enlarges font size;
4. power button ends call—uses the Power button to end calls;
5. auto-rotate screen—auto-rotates screen orientation;
6. speak passwords—speaks out your passwords;
7. accessibility shortcut—instantly accesses accessibility features with a button and touch combo;
8. text-to-speech—sets text-to-speech output;
9. touch and hold delay—adjusts touch and hold delay;
10. enhance web accessibility—installs scripts from Google to make the Web more accessible.

Apple's iOS 6 accessibility features are grouped according to disability: Vision:

1. VoiceOver—provides voice feedback and notification through gestures;
2. zoom—increases text size;
3. large Text—increases text size for Mail, Contacts, Calendars, Messages and Notes;
4. invert colours—inverts colours for less eye strain while reading text;
5. speak selection—text-to-speech output;
6. speak auto-text—speaks out auto-corrections and auto-capitalisations while typing.

Deaf and hard of hearing: Hearing Aids connects your device to supported hearing aids:

1. LED Flash for Alerts (on iPhone 5)—flashes the LED flash when receiving new alerts;
2. mono Audio—enables mono audio and adjustment of sound balance between the left and right channels.

Learning or physical and motor disabilities:

1. guided access—keeps the device in one app and control which features are available; triple tap Home button in the app you want to use;
2. assistive touch—assists you if you have difficulty touching the screen or if you need an adaptive accessory;
3. home-click speed—adjusts the speed for tapping the Home button to enable double and triple-click Home;
4. triple-click home—triple tap the Home button to access enabled accessibility features (VoiceOver, Invert Colours, Zoom, and AssistiveTouch).

Its large, crisp display (and also its larger virtual keyboard) helped Apple's iPad create the expanding tablet segment between smartphones and ultrabook computers. In 2012, iOS controlled an estimated 53.8 % of the tablet market, Android 42.7 % and Windows 2.9 % (mobiThinking 2013). This product segment also includes dedicated e-book readers including Kindle and Nook. Text-to-Speech (TTS) apps for tablets have made these devices even more attractive for those with vision and/or language disabilities (Royal National Institute of Blind People 2013). The iPad's compatibility with communication apps, such as Proloquo2Go, has made it a popular choice for alternative and augmentative communication (AAC) devices for children with speech communication disabilities (Disability Sanctuary 2013).

3 Wireless Access and Independent Living

The digital divide can be also a social divide—without access to mainstream consumer ICT, one is literally and figuratively not part of the conversation. For people with disabilities, who already face considerable obstacles to social and economic participation, access to wireless technologies is especially critical.

People with visual or hearing limitations have traditionally relied on a variety of assistive technologies, including hearing aids, magnifiers, currency identifiers, e-book readers, text telephones (TTY's), and Braille displays. Today's customisable electronic platforms with the ability to add downloadable applications ('apps') enable users to carry some of these assistive technologies right in their smartphones. Some smartphone features, such as GPS, offer services that have not previously been available through stand-alone assistive technologies.

4 Discovering Access Issues Through In-Person Research

The Wireless RERC conducts in-person user research through focus groups and product testing. From January through May 2013, the Wireless RERC conducted one take-home usability test (followed by a focus group with the test participants) plus two additional sets of focus groups in collaboration with partners in the wireless industry.

The take-home usability test included only people with visual impairment (blind and low vision), and focused on the accessibility of two specific smartphone models running the Android 4.2 operating system. The first set of focus groups included one group with visual impairment and one group with hearing loss (deaf and hard of hearing). The second set of focus groups included one group each for visual impairment, hearing loss and dexterity impairment. These last two sets of focus groups explored the out-of-the-box experience of smartphone users, regardless of the device owned by the participants.

The 44 participants with visual impairment and hearing loss in these several in-person studies varied in age (18–70; mean age of 44.4 years), race and ethnicity (22 White/Caucasians and 22 Black/African American) and gender (22 males and 22 females).

Following are five dominant themes identified from the Wireless RERC's 2013 focus groups and mobile handset testing among deaf, hard of hearing, blind and low-vision wireless customers. Overall, it was observed that most blind participants owned iPhones and held strong preferences for these devices. Deaf participants, on the other hand, displayed a greater tendency to own other types of smartphones, including Android-based and Blackberry devices.

4.1 The Out-of-the-Box Experience

Like many customers, people with visual or hearing limitations often find themselves on their own when choosing a new device and learning to use it. Some turn to online support or friends for help, but many appreciate the satisfaction of 'figuring it out on my own'. Video tutorials hold the potential for another source of support, but captioning and descriptive audio are generally less than useful for either blind or deaf customers. The Apple ecosystem is focused on a small family of similar devices and offers considerable help from the user community through online user groups. The broad diversity of Android devices offers more choice in device design, e.g. tactile keyboards. At the same time, this can make the Android ecosystem more difficult to navigate for customers with visual or hearing limitations.

4.2 Accessibility Settings

As outlined in Sect. 2, both Android and Apple enable users to tailor their devices to their own unique abilities, limitations and preferences through a menu of accessibility choices generally located in the 'Settings' menu. Since these settings can be crucial to the usability of the device, they must be easily discoverable and operable by the first-time user.

4.3 Incoming Calls, Messages and Other Alerts

Blind users appreciate phone features such as assignable ring tones and audio caller ID. Deaf and hard of hearing customers, on the other hand, suggest a flashing alert for incoming mail or calls and the option to assign different cadences or rhythms to different types of communications, i.e. text messages, phone calls, emergency alerts. Both groups appreciate the vibration feature in their device to alert them of incoming calls, emails, text messages, etc. To be useful, however, the vibration must be noticeable even when the device is clipped to a belt or held in a purse. Availability of these features, and ease in setting them up, clearly affects satisfaction with a given device.

4.4 Screen Readers

Screen readers are applications that add spoken, tonal and tactile cues to the visual displays of touchscreen devices, making them much more usable for blind and low-vision customers. Android devices use an application called TalkBack, while

Apple's solution is called VoiceOver. Blind and low-vision testers at the Wireless RERC have noted that, to be truly useful, screen readers must be easy to activate and convenient to use. And in situations where the feature is not desired, e.g. a concert, it must be easy and quick to disable.

4.5 Voice Recognition

As the power of mobile wireless technology increases, visual or hearing limitations no longer prevent customers from becoming 'power users'. Voice recognition, or speech-to-text, facilitating hands-free and eyes-free control, contributes to these users' proficiency. Both Android and iOS devices incorporate this feature. Personal assistant apps now also provide customers with the tools to communicate, navigate, access information and conduct transactions online. At the time of this writing, Google Now and Siri have their own individual strengths, and a clear choice must be a personal one. One important consideration might be that Google Now is currently available on both iOS and Android, while Siri is an iOS-only feature. This is one of the frontiers of mobile wireless technology where Android and iOS devices will continue to compete for customers with and without disabilities.

5 Findings of the Wireless RERC's Survey of User Needs (SUN)

Wireless RERC focus group and user testing research to date suggests dominance of Apple's iOS among blind users, but not among deaf users. These results from our qualitative research can be tested quantitatively by analysing response data of these same two disability groups in the Wireless RERC's SUN. Additionally, analysis of survey responses from blind and deaf respondents can reveal how effectively various operating systems are meeting the needs of these customers with disabilities.

The SUN was originally launched in 2002, to ensure that RERC research, development and training activities are guided by users themselves. This unique, nationwide survey on wireless technology use by people with all types of disabilities has come to be an important reference for people with disabilities, disability advocates, regulators, the wireless industry and other researchers. SUN data are regularly utilised by the wireless industry and government to guide their initiatives.

We invite the public to complete the SUN and share how wireless technology affects daily life, and how it could be improved. Data presented here are based on a non-random sample. The survey is promoted as broadly as possible through

Table 1 Respondent demographics	Blind	Deaf
Age mean	52	55
Age range	19–74	19–85
Female–male %	57–43	54–46

convenience sampling techniques, with special effort toward reaching under-represented groups. Sampling bias is partially corrected by weighting the response data by household income compiled in the 2011 American Community Survey (ACS) of the U.S. population of people with disabilities. ACS microdata are provided by the Integrated Public Use Microdata Series (IPUMS-USA) project at the University of Minnesota (Ruggles et al. 2010). Weighting the SUN response data by income helps to mitigate potential biases introduced by the convenience sampling approach. Household income was chosen for the sample weight because it is directly related to smartphone ownership and experience: as income rises smartphone ownership also rises (Morris et al. forthcoming). Household income also is strongly correlated with education level in the ACS sample.

The SUN has been updated over the years to keep up with the rapid pace of change in consumer technology. The results presented in this paper were compiled from the respondents to the Wireless RERC's fourth Survey of User Needs (SUN 4), launched in the fall of 2012. Of the 1,348 respondents, 85 are blind and 122 are deaf. One respondent who reported being both blind and deaf was excluded from this analysis in order to focus exclusively on the comparison between blind and deaf users.

Table 1 shows select demographics for blind and deaf participants in the survey. The relatively high values for mean age result to a substantial degree from the exclusion of minors under age 18 from the sampling.

5.1 Wireless Use and Operating Systems of Blind and Deaf Consumers

Table 2 illustrates the percentage of blind and deaf respondents who reported using mobile phones and tablets, and also who reported having wireline service in their homes. Notably, deaf users are more likely to use smartphones and tablets than blind users. The data in Table 2 also show that a majority of blind users own a smartphone.

Selecting just the respondents who said they had either 'no wireless device', 'basic phone' or 'smartphone' produces a gamma coefficient of 0.515 at the $p < 0.001$ significance level, indicating a very strong and significant relationship between disability type (blindness and deafness) and level of sophistication of mobile device used. Accordingly, deaf respondents are significantly more likely to

Table 2 'If you own or use a cell phone or tablet, what kind do you use?' (check all that apply)

	No wireless device %	Basic cell phone %	Smart phone %	Tablet %	Wireline %
Blind	8	30	54	17	83
Deaf	6	12	69	50	53

'Is there at least one telephone inside your home that is currently working and NOT a cellphone?'

Table 3 'If you own or use a SMARTPHONE, what kind do you have?' (check all that apply)

	Android %	Apple iOS %	Blackberry %	Windows %	WebOS %
Blind	18	86	6	8	4
Deaf	34	53	15	2	0

own more sophisticated wireless devices than blind respondents. Deaf respondents are also far less likely than blind respondents to have a functioning wireline phone in their homes.

These results likely reflect the need of deaf users for access to efficient text messaging. Most wireline phones do not provide captioning/text writing capabilities (although TTY is still available in the United States, and telephone service providers like Sprint offer caption-enabled wireline devices and service). Most simple (non-smart) mobile wireless phones have text messaging capabilities that are supported by wireless service providers, but they are more cumbersome to use for composing and reading text messages. Deaf consumers embraced the original T-Mobile Sidekick and its successors, which offered a slide-out QWERTY keyboard, after it was introduced in the United States in October 2002, and anecdotal evidence (and survey data presented in Table 3) suggests they also embraced the classic Blackberry phone with the physical QWERTY keyboard.

Greater smartphone ownership rates by deaf respondents compared to blind respondents might also result from greater ease of use of the numerous vision-based features on smartphones as experienced by deaf users relative to blind users. Recent accessibility innovations for blind users such as Siri and VoiceOver on iOS, and Google Now and Explore by Touch on Android do not fully overcome the visual access challenges that blind users experience, which non-blind deaf users do not experience. Simple mobile phones with numeric keypads and raised physical keys and a nub on the number '5' key for orientation, on the other hand, can be accessed effectively by blind users.

Table 3 shows the use of Android, Apple iOS, Blackberry, Windows and WebOS operating systems among blind and deaf smartphone users. Together, Android and iOS serve the majority of these customers, while 17–18 % of each group uses Blackberry, Windows or WebOS. Notable in these results is the overwhelming use of iPhones by blind respondents, and the much higher rate of use of Android and Blackberry by deaf respondents.

Comparing just the use of iPhones and Android smartphones among blind and deaf respondents produces a gamma coefficient of 0.589 at the $p < 0.01$

Table 4 How satisfied are you with your primary wireless device?

	Blind		Deaf	
	Android ($n = 5$)[a] %	iOS ($n = 29$) %	Android ($n = 22$) %	iOS ($n = 40$) %
Very satisfied	42	77	35	62
Somewhat satisfied	0	23	48	38
Neither satisfied or dissatisfied	39	0	9	0
Somewhat dissatisfied	0	0	7	0
Very dissatisfied	19	0	0	0

Primary device
[a] Percentages reflect impact of weighting the sample by income, and therefore vary somewhat from expected values calculated by dividing the number of cases fitting selected criteria by the total number of cases

significance level, indicating that blind respondents are significantly more likely to own iPhones than Android smartphones.

Table 4 demonstrates differences in satisfaction between blind and deaf users of Android smartphones and iPhones. All blind and deaf iPhone users reported being either 'Somewhat satisfied' or 'Very satisfied' with their wireless devices. Among Android customers, 83 % of deaf users, but only 42 % of blind users were 'Somewhat satisfied' or 'Very satisfied' with their devices. It should be noted that there were very few blind users of Android smartphones, making this analysis less reliable.

Table 5 shows similar differences between Android smartphone and iPhone users: 96 % of deaf users and 85 % of blind iPhone users described their devices as 'Easy' or 'Very easy' to use. Among Android smartphone owners, 80 % of deaf users and 42 % of blind users described their primary devices as 'Easy' or 'Very easy' to use. Again, the number of blind android smartphone users is low, making analysis less reliable.

6 Discussion

Data presented from the Wireless RERC's SUN suggest that Apple's iPhone is currently better than Android-based smartphones at meeting the needs of blind consumers. Differences in adoption rates, ease of use and satisfaction between the two operating systems are less pronounced among deaf users.

These results likely reflect in part the experiences of customers with devices that have been on the market for more than the past year, during which time the accessibility features and capabilities of the Android operating system have expanded considerably.

The survey results also reflect in part how people with disabilities (and perhaps those without disabilities) choose their mobile phones. Socially reinforced patterns of preferences and economic choices can develop among close-knit communities

Table 5 How easy or hard is your wireless device to use?

	Blind		Deaf	
	Android $(n = 5)^{a}$ %	iOS $(n = 29)$ %	Android $(n = 22)$ %	iOS $(n = 40)$ %
Very easy	42	56	22	68
Easy	0	29	58	28
Somewhat hard	39	13	20	4
Hard	19	2	0	0
Cannot use it without help	0	0	0	0

Primary device

[a] Percentages reflect impact of weighting the sample by income, and therefore vary somewhat from expected values calculated by dividing the number of cases fitting selected criteria by the total number of cases

bound together by shared experiences and challenges that are distinct from those of the general population. Such patterns seem to have developed among blind and deaf consumers, and others with disabilities. The Wireless RERC's survey research shows that when making their mobile technology purchase decisions, people with disabilities rely most commonly on recommendations from friends, family and others in their personal network (Wireless RERC 2013). Blind and deaf consumers are among the most likely to seek out recommendations from their personal networks. Other sources of product information such as online consumer information sources, websites of service providers and device manufacturers, and sales personnel are all much less frequently consulted when making a purchase decision.

These patterns have been highlighted by participants in the Wireless RERC's focus group and user testing research. Disappointed by limited understanding of disability by customer service personnel working for handset manufacturers and wireless carriers, blind and deaf consumers have reported that they rely primarily on peers for help in choosing and using their new device.

The Wireless RERC's focus group and user testing research also revealed the substantial challenges that blind and deaf consumers experience when switching between Android, iOS or other platforms. A successful out-of-the-box experience requires personal commitment and access to assistance. Particularly for blind consumers, switching to a new device and operating system represents a considerable learning curve affecting satisfaction and ease of use, at least in the short term.

For these customers, Apple's comparatively well-defined ecosystem makes locating peer support (and 'figuring out' a new device) an easier task. In contrast, the rich landscape of device choices available to Android customers is a mixed blessing for those with disabilities. Hardware and software variations across platforms and carriers result in unique accessibility characteristics for each device. Usefulness and usability of accessibility features and assistive applications also vary across devices and carriers. Discovering and understanding all the available choices can be challenging. Nevertheless, for both Android- and iOS-driven devices, effective technical assistance informed by genuine disability awareness is an emerging priority for the wireless industry.

The accelerating pace of wireless technology development, and the importance of the large and growing market of customers with disabilities, guarantees that healthy competition in this arena will continue. This is a promising sign for those with disabilities, as well as the rest of the population, who are likely to live long enough to experience age-related loss of function.

Acknowledgments The Rehabilitation Engineering Research Center for Wireless Technologies is funded by the National Institute on Disability and Rehabilitation Research of the U.S. Department of Education, grant #H133E110002. The opinions contained in this document are those of the grantee and do not necessarily reflect those of the U.S. Department of Education.

References

Disability Sanctuary (2013) AssistiveWare talks autism apps, price challenges and iOS versus Android. Available at: www.disabilitysanctuary.com/forum/index.php?threads/assistiveware-talks-autism-apps-price-challenges-and-ios-versus-android.8554/ (Accessed on 29 October 2013)

mobiThinking (2013) Global mobile statistics 2013 Part A: mobile subscribers; handset market share; mobile operators. Available at: http://mobithinking.com/mobile-marketing-tools/latest-mobile-stats/a#mobiletablet (Accessed on 29 October 2013)

Morris J, Mueller J, Jones M, Lippincot B (forthcoming) Wireless technology use and disability: Results from a national survey. Int J Technol Persons Disabil. Northridge California, California State University at Northridge

Ruggles S, Alexander JT, Genadek K, Goeken R, Schroeder MB et al. (2010) Integrated public use microdata series: version 5.0 (machine-readable database). University of Minnesota, Minneapolis

Royal National Institute of Blind People (2013) Accessibility of eBooks. Available at: www.rnib.org.uk/livingwithsightloss/reading/how/ebooks/accessibility/Pages/ebook-accessibility.aspx (Accessed on 29 October 2013)

Smith K (2012) Here's why Android is better than Apple's new operating system for the iPhone. Available at: www.businessinsider.com/google-android-jellybean-versus-apple-ios-6-2012-9?op=1#ixzz2Wl7s3R5V (Accessed on 29 October 2013)

US Government Printing Office (2010) Twenty-First Century Communications and Video Accessibility Act of 2010. Available at: www.gpo.gov/fdsys/pkg/PLAW-111publ260/pdf/PLAW-111publ260.pdf (Accessed on 29 October 2013)

Wireless RERC (2013) SUNspot—adults with disabilities and sources of wireless accessibility information. Available at: http://wirelessrerc.org/content/publications/2013-sunspot-number-06-adults-disabilities-and-sources-wireless-accessibility (Accessed on 29 October 2013)

Ybanez (2013) Android 4.2 Jelly Bean vs. Apple iOS 6.1—which is the sweeter treat? Android Authority. Available at: www.androidauthority.com/android-4-2-1-jelly-bean-vs-ios-6-1-comparison-154219/ (Accessed on 29 October 2013)

Part III
Reconciling Usability, Accessibility and Inclusive Design

Preliminary Results in the Understanding of Accessibility Challenges in Computer Gaming for the Visually Impaired

J. Chakraborty, J. Hritz and J. Dehlinger

1 Introduction

Today's gamers are confronted by a plethora of game choices ranging from simple, two-dimensional board games to highly sophisticated and visually intensive video games. The gaming industry is experiencing a period of growth and innovation through content delivery (e.g. mobile devices, social media platforms, etc.) and user interface/interaction modes (e.g. motion capture/detection, haptic feedback, online game play, etc.). Video games have been found to be beneficial and should not be viewed only as vehicles for entertainment. Research has found the video games promote a healthy sense of competition (Vorderer et al. 2003; Velleman et al. 2004), help develop social skills among special needs groups (Griffiths 2002), reduce differences in spatial recognition between genders (Feng et al. 2007), have health benefits in clinical settings (Griffiths 2005), etc. However, the recent growth and innovation in user interfaces and interaction modes in the gaming industry has outpaced the software engineering work needed to understand the accessibility challenges faced by visually impaired users and develop requirements to give visually impaired users equal opportunities to access video games.

An important step in developing video games for users who are visually impaired is to do a usability evaluation and understand the accessibility challenges faced by this community and the user interface requirements needed to better accommodate them. Understanding how traditional games designed for sighted individuals can be provisioned with the interface/interaction modes specific to visually impaired users will enable and encourage game developers to design for universal usability/accessibility. As a start, this work seeks to better understand how visually impaired gamers interact with traditional games, their accessibility challenges and the necessary software engineering non-functional requirements

J. Chakraborty (✉) · J. Hritz · J. Dehlinger
Universal Usability Laboratory, Department of Computer and Information Sciences,
Towson University, Towson, USA
e-mail: jchakraborty@towson.edu

P. M. Langdon et al. (eds.), *Inclusive Designing*, DOI: 10.1007/978-3-319-05095-9_8,   83
© Springer International Publishing Switzerland 2014

when designing and developing inclusive video games for visually impaired gamers. Specifically, the contributions of this paper include:

- An exploratory usability study to understand the accessibility challenges of visually impaired users in their interaction with a developed game.
- An observational analysis and preliminary results of the exploratory usability study derived from six visually impaired gamers playing a developed game over 30 min evaluation periods.

This work is part of a larger effort to investigate the challenges encountered by visually impaired gamers when interacting with standard, PC-based games in order to develop a software engineering understanding of the usability/accessibility requirements to give the visually impaired equal opportunities to access video games, and the user interface design implications for future iterations of video game developers.

The remainder of this paper is as follows. Section 2 provides a brief summary of the literature in gaming accessibility research challenges and the limited empirical work that has been carried out in the field. Section 3 outlines the experimental design carried out for usability testing a typical PC-based arcade game with six visually impaired gamers. Section 4 details the preliminary results derived from an observational analysis of the usability testing, and, finally, Sect. 5 concludes with a discussion of the implications for future iterations of video game developers and of our future research directions in analysing the usability study data.

2 Background

Research in the field of computer gaming usability and accessibility is lacking. Very few empirical studies have been carried out to understand the software design requirements of visually impaired gamers. Bierre et al. (2005) illustrate this point clearly in their position paper. Through their work at the International Game Developers Association (IGDA), they have highlighted a need for universal accessibility for gamers. Subsequent research has focused on navigation issues, including Trewin et al. (2009) who carried out an empirical study of 3D virtual world gaming by infusing an extensive set of accessibility features into the virtual world game PowerUp. Their findings revealed a need for additional accessibility features such as navigation and audio support. Westin (2004) developed a real-time, 3D graphic game focusing on gaming accessibility for the blind. The study focus was on sound interface and its relation to the 3D graphics world. Sanchez et al. (2009) developed an Audio-Based Environments Simulator (ABES) to assist blind users to navigate through a virtual representation of space. However, this software has not been tested within a gaming environment where end users have to constantly change their game strategy.

Velleman et al. (2004) designed a curriculum to develop accessible games for the blind. The authors point out that through user testing, several accessible games have been developed. However, their work does not provide empirical evidence of user requirements for blind gamers.

Yuan et al. (2011) developed an innovative approach to gaming accessibility for the blind. The researchers developed a glove that would transform visual cues from the game Guitar Hero into haptic feedback to the tips of the user's fingers and found that visually impaired users enjoyed the increased accessibility of the game.

Carvalho et al. (2012) carried out a research study of mobile gaming for the blind. In it, the researchers presented an audio-based puzzle game designed using an iterative participatory technique to 13 blind participants. The findings illustrate that the blind gaming community can significantly benefit from these innovative interactive design techniques.

3 Experiment Design

This work seeks to better understand how visually impaired gamers interact with traditional games, their accessibility challenges and the necessary software engineering non-functional requirements for designing and developing video games inclusive for visually impaired gamers. To understand the challenges of visually impaired gamers, we developed a PC version of a 1980s arcade style video game, entitled Ninja Cactus (Fig. 1). Ninja Cactus incorporates joystick controls and sound effects and was designed as a traditional, 2D, arcade game (i.e. not designed specifically for blind users). We gathered gamer experience data from six visually impaired users and designed a pre-game questionnaire to gather video game playing background data about each subject. We then created a list of visual cues to look for during game playing. Finally, we created a post-game questionnaire to gather end user data.

3.1 Game Summary

The premise of Ninja Cactus is based on *Suparutan X*, a game developed in 1984 by Irem, and published in the United States by Data East as Kung Fu Master. One of the first brawler games, Kung Fu Master features a protagonist Kung Fu fighter who must trek across multiple two-dimensional levels in order to save his girlfriend. Although the main character can jump, the game plays out over a flat horizontal baseline without any ground level change. Throughout each level, various enemies spawn and attempt to attack the player. These attacks can be as simple as running towards the character, to more advanced attacks such as using a projectile. To beat this game, the player must fight through these levels using two basic attacks, a kick and punch, coupled with three moves, standing, crouching and jumping.

Fig. 1 Screen shot of Ninja Cactus video game in play

Similar to Kung Fu Master, Ninja Cactus features a character that must fight its way through a unidirectional level, where various enemies spawn to try and attack. The cactus must travel across this level populated by five types of unique enemies, who act in various ways. Borrowing from other games from the Kung Fu era, Ninja Cactus can shoot a projectile similar to Mega Man, a character from a popular platform game series developed by Capcom. This 'fireball' along with a kick and punch are the tools the player must use to negotiate the level. Just as the player must utilise the various attacks to move forward in Kung Fu Master as each enemy can only be defeated with a subset of the available attacks, so too must the player fight forward in Ninja Cactus with similar game mechanics. Ninja Cactus rewards the player who can learn how each enemy behaves and execute an appropriate attack to defeat these enemies.

3.2 Game Player Rewards

During the game, player rewards come in the form of points. Each defeated enemy is worth a different amount of points according to the relative difficulty of defeating him. Enemies with simple behaviours are worth less than those enemies with more complex attack loops. Some enemies possess more hit points, requiring the player to hit them with more than one attack to defeat, and award the player more points than those with less. Points are also awarded for moving forward in the game, which is left in the playing field, and they are also taken away if the player retreats. Upon reaching the final boss the player receives points, and like the other enemies, points are awarded for the boss' defeat, although much more generously as this final enemy requires a player to learn a more complex set of attack sequences and discover when the boss is vulnerable. Once the player defeats the end boss or exhausts three of the cactus' lives the game is over; it switches over to a highpoints list highlighting the player's score within this list if it is high enough.

3.3 Game Movement

The game's character movement is limited to the two-dimensional axes. The character can jump, move left and right using the toggle keys on the keyboard; however, as it typically stays stationary on the horizontal axis, left and right movement must be simulated. Background scrolling helps to create the sense that the cactus is moving through the level.

To reinforce a sense of movement, there are two other features to this game. First is the parallax background scrolling feature, where the foreground scrolls faster than the background, which also simulates a third dimension, giving the playing field some depth. The second visual cue to aid in the sense of motion is the movement of the character relative to both the enemies and the clouds in the background. As the player moves Ninja Cactus, most of the enemies will move in relation to the character. Since these enemies are tied to the background, if standing still, they will appear to be stationary in respect to the background, appropriate for a character standing still.

3.4 Game Sound

The game incorporates several sound features. From the onset, the user hears the general game theme sound which is a constant tune, similar to the original Kung Fu Master theme. In addition, each attacking motion, a kick, a punch or a leap, by the Cactus Ninja results in a single sound. This sound is different from the ones emanating from an enemy character. When enemy attacks, using fireballs and knives, their actions also result in a single sound. As the enemy attacks are successful, the Ninja Cactus produces a sound signifying a successful enemy attack. The sounds have been incorporated into the gaming software and cannot be turned off manually.

3.5 Data Collection Instruments

Institutional Review Board (IRB) permission was sought, and subsequently obtained, to carry out data collection using subjects from the National Federation for the Blind offices in Baltimore, Maryland, US. Invitations were sent out by email to the blind community seeking participants for the study. Six random subjects chose to participate. After the objectives of the study were explained, each subject was interviewed using the pre-game questionnaire. This was designed to collect background information, such as gaming experience, about each subject. The subjects were then given time to get used to the keyboard game controls and to familiarise themselves with the video game with the help of one of the investigators. They could ask as many questions as desired to familiarise themselves with the interface. Great care was taken to ensure that engagements with the subject would not bias their views of the study. The subject then played the video game for a short time on a laptop PC while another investigator recorded their observations and after a brief time, the subject was then interviewed about their experience.

Part of our study was to observe the behaviours of the participants as they engaged with the various aspects of the game in its existing state. With that aim, we observed user behaviours using the following protocol:

- Note the interactive moves the subject displays before actual start of play.
- Observe emotional cues and outputs the subject displays before the start of play.
- As the subject plays the game and the game character 'dies', evaluate the subject's understanding of the controls, objectives of the game and sounds by recording the subject's expression.
- As character 'dies', ask the subject about the controls: 'Good; average; bad'.
- As character 'dies', ask the subject whether the game's sonic feedback is helpful: 'Good; average; bad'.
- Note whether the subject developed a gaming strategy.

4 Preliminary Results

Once the familiarisation period with the game and keyboard controls was completed, limited help was offered to the subjects during the data collection process to insure against bias. The observations were recorded by two investigators using pen and paper and compiled together. Every effort was made to insure against bias.

The four male and two female subjects were aged from 18 to 40. Of the six subjects, two had identified themselves as being partially sighted. The subjects had indicated that they had engaged in video gaming at various levels of expertise for 8–10 years. The six participants had various levels of computer proficiency.

4.1 Observations

Subject one was a male, aged 35–40. He had a master's degree and was a working professional in the computer industry. This right-handed subject had been a gamer for more than 10 years and was very familiar with a non-Braille keyboard. He was initially very enthusiastic as the game was being explained to him. However, this enthusiasm was visibly waning as the subject played the game for a few minutes. He incorporated a trial-and-error approach to playing the game since the objectives were not very clear to him. He became increasingly frustrated as the game did not have sounds to differentiate between the directions of oncoming enemies. At one stage, he had no idea if his game character was still alive. The subject found the sound layout confusing and complained about the background noise. He suggested that the game could be enhanced if each toggle key could be associated with a sound to help him determine what action was being carried out. Overall, the subject was not impressed by the limited accessibility of the game.

Subject two was a male aged 25–30. He had completed his bachelor's degree and was a part-time worker in a field using computers. This right-handed subject had been a gamer for approximately 7 years. He was patient and inquisitive when the initial game was being explained to him, but was a very tentative user, who took some time to get acclimatised to the gaming environment. From the beginning, the subject was unsure and uncomfortable with the keyboard layout and the sounds. He never fully understood the gaming premise and was not successful in completing a level. His sole strategy was to get away from the enemy as fast as possible and to that effect, he incorporated a 'button mashing' strategy, where he would press all the buttons he was informed about at the same time. This was not very successful. The subject complained about the lack of directional sounds associated with the movement of his character and the enemies, and was unhappy with the overall directional orientation and the sounds incorporated in the game.

Subject three was a male in his late 20s. He had completed his master's degree and was a working professional in a mid-level management capacity at a large organisation and was very familiar with using computers. This right-handed subject had been a gamer for approximately 12 years, and as a keen gaming enthusiast he had a very hands-on approach to learning. As the game was being explained to him, he moved his hands around to get a spatial sense of the game environment. He understood the keyboard layout very well, but he had a hard time relating the keys to the associated movements on the screen. The subject could not distinguish between the sounds but was willing to incorporate several running, jumping and attacking strategies to beat the game. However, he began to get frustrated at his inability to master the game controls as the game continued, was unhappy with the sound effects in the game and rated the control layouts as average.

Subject four was a female aged between 18 and 22. She was a college student pursuing a degree in Liberal Arts who identified herself as partially sighted. This right-handed subject had been a gamer for approximately 5 years. Enthusiastic in the beginning, she became frustrated very quickly: she was unfamiliar with a non-Braille keyboard and was constantly re-positioning her fingers. She was in deep concentration to grasp the game concept and tried several ways to beat the game. The subject was constantly in attack mode without moving towards any game objectives and complained about the lack of sound directions to guide her. She was disoriented in the game environment and unhappy with the controls, suggesting a simpler keyboard layout for the blind.

Subject five was a male aged between 18 and 22. He was a college student pursuing a bachelor's degree in Mathematics who self-identified himself as partially sighted. This right-handed subject had been a gamer for approximately 5 years. Being very familiar with a non-Braille keyboard, he grasped the controls of the game very quickly and went through several levels of the game incorporating most of the movements comfortably. The subject was able to differentiate between the enemy sounds and that of his character and would engage the enemy successfully. Towards the end of the game the subject was pleasantly surprised to discover one additional movement, the leap. He had achieved the highest score and was pleased with nearly all aspects of the game.

Subject six was a female aged between 18 and 22. She was a college student pursuing a bachelor's degree but she had not identified a major. This left- handed subject had been a gamer for less than a year, so was not comfortable with a right-handed non-Braille keyboard and struggled greatly with the controls. The subject did not understand the premise of the game as she was coming to grips with the controls: she tried to press a few buttons to tentatively get some movement and develop a strategy of press and see what happens. She was not happy with any aspect of the game or the controls.

4.2 Discussion

All the participants were computer users with more than 5 years experience. Each participant had some college education or higher and several years of gaming experience. They were all given similar instructions and allocated similar amounts of time to familiarise themselves with the game controls. Every effort was made to avoid bias.

Our findings were consistent with the literature. All the participants were unhappy with the accessibility limits of the gaming interface. Every participant felt confused by the sound effects. Some participants were quite handicapped by the non-Braille keyboard. The objectives of the game were not clearly understood by most. The lack of a clear set of guidelines and accessible controls resulted in frustration and user dissatisfaction.

The most common feedback from the participants related to the lack of directional sound to indicate movement of the main character and the in-coming enemies. Other complaints were directed at the lack of sound from the game menu. It was unclear if the participants' performances with the game would have improved, had more time been allocated.

5 Future Research Directions

The preliminary results reported in this paper stem from an observational analysis carried out from the notes taken during the exploratory usability study performed with six visually impaired users playing a developed game over 30 min evaluation periods. In addition, the exploratory usability study included semi-closed pre- and post-game interviews. The interview questions beforehand were aimed at determining the users gaming experience/habits and modes of interaction that best suit visually impaired user interaction. The questions asked afterwards were specifically aimed at how traditional, PC-based arcade games, such as the one used in the usability study, can be augmented to better accommodate the modes of interaction for visually impaired gamers.

The long-term goal of this work is to better understand how visually impaired gamers interact with traditional games, their accessibility challenges and the necessary software engineering non-functional requirements to design and develop video games inclusive for visually impaired gamers. Following the preliminary results reported here, we plan to methodically analyse the qualitative data resulting from the semi-closed pre- and post-game interviews using an adaptation of Strauss and Corbin's (1990) grounded theory method (GTM). Specifically, our adaptation of GTM (Chakraborty and Dehlinger 2009; Chakraborty et al. 2012) systematically analyses qualitative text, such as the interview transcripts gathered in the usability study in this work, and develops software engineering artefacts based on user feedback and evaluation, in this case operationalisable, non-functional requirements that give visually impaired gamers the same opportunities as full-sighted gamers. The aim is to develop user interface/user interaction guidelines to give software engineers the knowledge to develop universally usable games.

References

Bierre K, Chetwynd J, Ellis B, Hinn DM, Ludi S et al (2005) Game not over: accessibility issues in video games. In: Proceedings of the 3rd international conference on universal access in human-computer interaction

Carvalho J, Guerreiro T, Duarte L, Carriço L (2012) Audio-based puzzle gaming for blind people. In: Proceedings of the mobility accessibility workshop at MobileHCI, San Francisco, CA, US

Chakraborty S, Dehlinger J (2009) Applying the grounded theory method to derive enterprise system requirements. In: Proceedings 10th ACIS international conference on software engineering, artificial intelligences, networking and parallel/distributed computing, Daegu, Korea

Chakraborty S, Rosenkranz C, Dehlinger J (2012) A grounded theoretical and linguistic analysis approach for non-functional requirements analysis. In: Proceedings of the 2012 international conference on information systems, Orlando, FL, US

Feng J, Spence I, Pratt J (2007) Playing an action video game reduces gender differences in spatial cognition. Psychol Sci 18(10):850–855

Griffiths M (2002) The educational benefits of videogames. Educ Health 3:47–51

Griffiths M (2005) Video games and health: video gaming is safe for most players and can be useful in health care. Br Med J 331(7509):122

Sánchez J, Tadres A, Pascual-Leone A, Merabet L (2009) Blind children's navigation through gaming and associated brain plasticity. In: Proceedings of virtual rehabilitation 2009 international conference, Haifa, Israel

Strauss AL, Corbin JM (1990) Basics of qualitative research. Sage, Newbury Park

Trewin S, Laff M, Hanson V, Cavender A (2009) Exploring visual and motor accessibility in navigating a virtual world. ACM Trans Accessible Comput (TACCESS) 2(2):11

Velleman E, van Tol R, Huiberts S, Verwey H (2004) 3d shooting games, multimodal games, sound games and more working examples of the future of games for the blind. In: Proceedings of the 9th international conference on computers helping people with special needs, Paris, France

Vorderer P, Hartmann T, Klimmt C (2003) Explaining the enjoyment of playing video games: the role of competition. In: Proceedings of the 2nd international conference on entertainment computing, Pittsburgh, PA, US

Westin T (2004) Game accessibility case study: Terraformers–a real-time 3D graphic game. In: Proceedings of the 5th international conference on disability, virtual reality and associated technologies, Oxford, UK

Yuan B, Folmer E, Harris FC Jr (2011) Game accessibility: a survey. Univ Access Inf Soc 10(1):81–100

Investigating Accessibility to Achieve Inclusive Environments: The Spatial Experience of Disability at a University Precinct in Lisbon

T. Heitor, V. Medeiros, R. Nascimento and A. Tomé

1 Introduction

According to the universal design paradigm, accessibility in the built environment indicates the degree to which any space to be used by people is reachable by someone with a permanent or temporary impairment (Levine 2003). A disadvantage (handicap) is not just a characteristic of people with disabilities. It can occur with anyone whenever the demands of the environment exceed their capabilities. The options are to raise people's capabilities or reduce the demands of the built environment. Intervention at the level of people's capabilities is possible and necessary. However, without intervening at the level of the built environment, there will always be people whose capabilities fall short of the demands made upon them. The idea is that through a deliberate design process that focuses on the needs of all users, especially including persons with all kinds of disabilities, most of the built environment can be improved for everyone, greatly expanding the range of users.

A failure of accessibility would become a barrier that effectively isolates many groups of people, preventing them from meeting others and holding them back from participation in social, educational and working activities. When looking at a

T. Heitor (✉)
Instituto de Engenharia de Estruturas, Território e Construção, Instituto Superior Técnico, Department of Civil Engineering, Architecture and Georesources, University of Lisbon, Lisbon, Portugal
e-mail: teresa.heitor@ist.utl.pt

V. Medeiros
Faculty of Architecture and Urbanism, University of Brasilia, Brasilia, Brazil

R. Nascimento
Instituto Superior Técnico, Department of Civil Engineering, Architecture and Georesources, University of Lisbon, Libson, Portugal

A. Tomé
Instituto de Engenharia de Estruturas, Território e Construção, Instituto Superior Técnico, Department of Civil Engineering, Architecture and Georesources, University of Lisbon, Lisbon, Portugal

P. M. Langdon et al. (eds.), *Inclusive Designing*, DOI: 10.1007/978-3-319-05095-9_9,   93
© Springer International Publishing Switzerland 2014

given route from the perspective of disabled people, it is critical to assess all movement barriers along the way to get an idea of how hard it will be for them to use this route, and whether they will prefer another (Imrie and Hall 2001).

There is a growing body of literature on accessibility as well as a considerable number of building codes and regulations, barrier-free standards and generic design guidance governing the accessibility of the built environment. They are, in general, limited in scope and complaint-based rather than proactive. They exist as a disembodied dataset of definitions, prescriptions, rules and tables, evolved through a process with a strong internal logic, but ignoring the 'geography of accessibility' (Hanson 2004) and the consequences of territorial and mobility-related needs of disabled people, in particular regarding the demand for excessive physical efforts and time consumption (Wagener and van der Spek 2006; Oliver 2008).

This paper is based on a study developed within the framework of IN_LEARNING research project (Nascimento 2012), using the IST campus in Lisbon as a case study. It has an exploratory perspective and intends to discuss the configurational variables, investigated by means of the theory of the social logic of space (Hillier and Hanson 1984), in studies about universal design and accessibility. It aims to analyse accessibility and the social and spatial dimensions of the experience of disabled individuals, taking into account the development of a mapping profile and the premises for an indicator to evaluate the level of effort (LoE).

2 IST Campus: A Scenario of Growth

The IST Campus, completed in 1937, was the first university campus to be built in Portugal. It is located in Lisbon and occupies a sloping site of about 10.4 ha within the inner city area, which defines an entire urban block.

The campus concept, focused on distinctiveness, is based on an enclosed territory surrounded by walls, which take the form of a semi-opaque boundary. It is composed of six building clusters including a canteen, a sports hall and a covered swimming pool. The layout follows a symmetrical spatial organisation defined by a central boulevard linking the main entrance to the front of the main building, which embodies the public face of the campus. This central boulevard is accessed by large and impressive staircases and provides indirect access to the different building clusters.

From late 1969 up to now, some of the existing buildings have been extended and new ones built, keeping the original campus area. Currently, the campus comprises a built area of $34,386 \text{ m}^2$ organised according to a faculty/related departmental structure and serves a population of about 12,000 users: 10,500 students, 1,000 faculty teachers and researchers and 500 administrative workers (NEP 2012).

The campus is accessible by a total of five entrances. Two of them are exclusively for pedestrians and are reliant on wide and high staircases without handrails. Three lateral entrances provide access by car. Nevertheless cars do not directly access all the building entrances. Parking provision is concentrated around the central building and in the surroundings of the campus.

It is well recognised by the academic administration that people with disabilities face multiple physical barriers to accessing the site and moving within it. The main 'black spots' (i.e. causes of difficult access) include car parking on sidewalks; the stairs are too wide and have no handrail, issues with the management and maintenance of the space as delivery platforms, the state of the cladding and vegetation. The existing routes have multiple barriers and it cannot be forgotten that the existence of just one barrier in the continuity of a route can restrict the movement of pedestrians, including disabled people. The internal configuration is very complex, with many corners between walls, slopes and buildings that produce an excessively fragmented space. Moreover, the central boulevard slope and pavement is not suitable for wheelchairs, and accessible paths through the campus linking the different building clusters are not clearly identified.

3 Methodological Procedures

This study was developed by means of a mapping profile using both high- and low-tech spatial description techniques, namely space syntax (Hillier and Hanson 1984) synchronised with a conventional measured survey involving direct data capture and map interviews, i.e. an inquiry procedure combining walkthroughs with semi-structured interviews, documentary analysis and space-use observations. It operates by combining a macroanalysis of the campus with a microanalysis of accessibility conditions and architectural barriers focused on the definition of an accessibility indicator—LoE. It was built by Nascimento (2012) to get information about the excessive physical effort and time consumption demanded of disabled people when overcoming such barriers as: stairs, ramps, pathways and sidewalks.

The mapping profile was designed to record and structure information and to provide a spatial framework within which accessibility conditions could be quantitatively compared and monitored, as well as a simple way for project participants and other key stakeholders to communicate, explore accessibility improvements and prioritise areas of intervention.

The spatial description of the IST campus was carried out in two stages. The first one was focused on the syntactic description of the campus by means of visibility/isovists and axiality/convexity techniques (Turner et al. 2001) (Depthmap® software) applied to the external circulation network. The axial description considered two alternative maps: one describing the overall circulation system, including stairs and ramps, which were defined as single axis on the axial map, and the other one excluding stairs as well as all the areas that are only accessible via stairs, i.e. it showed areas without architectural barriers. This second axial map corresponds to the experience of those travelling in a wheelchair. Besides revealing the importance of access ramps, it shows the extent of very hard-to-reach areas within the campus.

The visibility description also considered two alternative maps: one including car parking areas and the other one excluding them in order to understand how much undisciplined car parking using pedestrian areas breaks the links within the Campus.

The second stage was based on research-based fieldwork, including: (1) the survey of all architectural barriers—60 stairs, 61 ramps, 113 pathways and side-walks; (2) exploratory interviews with users and accessibility experts, including therapists and analysts, to access architectural barriers and explore different per-spectives on overall accessibility on campus and (3) space-use observations. A total of 10 participants were recruited for interviews and walkthrough experiments on a voluntary basis: two visually impaired, two in wheel chairs, one with a motor handicap and another one with a temporary disability (broken foot) moving with canes and three without any disability. All of them were familiar with the campus. Accessibility experts assessed the campus conditions and identified failures in accessibility and barriers responsible that created architectural obstructions.

All the participants were invited for a walkthrough experiment within the campus, beginning from the outside of their own department building (main entrance) and ending at the Canteen building. Navigation performance was mea-sured over the following variables: (1) time to complete the task; (2) stops; (3) distance covered and (4) average speed. Each route was described and individually analysed, in comparison with the routes performed by the three participants without disability. This approach was intended to determine the effect barriers have on users' behaviour. Walkthrough tests were recorded on video and timed.

Interviews with the participants were carried out to obtain information on typical routes used for daily travel within the campus, physical/architectural bar-riers and social and attitudinal barriers due to negative perceptions or feelings, and positive aspects including the presence of friendly people and accessible building features. Physical barriers were classified according to the excess physical effort required to overcome them (low, medium or heavy).

Interviews also covered questions about past experiences, perceptions and social attitudes about disability and expectations for the future regarding acces-sibility within the campus.

4 Results and Discussion

4.1 Space Syntax Analysis

Space syntax (Hillier and Hanson 1984) is a set of theories and techniques for the analysis of spatial layouts, which can simulate the social implications of spatial configurations—at both the urban and the building scale—including patterns of movement i.e. navigation patterns. It utilises a toolbox capable of describing—representing, quantifying and comparing—spatial systems and the relative connectivity and integration of the spaces. Spaces are understood as voids (e.g. streets, squares), between built elements (e.g. walls, fences) and other impediments or obstructions that restrict pedestrian movement and/or the visual field.

Accordingly, the spatial system is described by means of the axial map, i.e. the least set of longest and straightest lines of sight and access, which covers all the circulation spaces. Axial maps produce two types of output: alphanumeric data, in the form of a graph which is the basis for deriving measures of a configuration's properties from the viewpoint of users' mobility conditions, and graphic data in the form of core maps that present an immediate understanding of the spatial pattern of the area.

When considering the spatial pattern, the spectrum colours range from red: higher values; to blue: lower values. Integration measures how many turns one has to make from a given axial line to reach all other axial lines in the network, using shortest paths. Theoretically, the integration measure shows the cognitive complexity of reaching a space, and is often argued to 'predict' pedestrian movement. It is argued that the easier it is to reach a space the more it is going to be used. Connectivity measures the number of immediate spaces that are directly connected to a space. Intelligibility is the correlation between connectivity and integration and describes what can be understood of the global relations of a space by an observer within that space.

Syntactic analysis was applied to identify mobility conditions for users' and to evaluate how easy is to navigate within the campus. The campus syntactic model (Fig. 1) shows that the most highly integrated spaces (i.e. those spaces that express the highest levels of physical accessibility within the overall spatial system) correspond with the central boulevard (main pedestrian access) and the integration core, i.e. the set of the most accessible spaces of the system defines a ring surrounding the main building, which is linked to both the main pedestrian and car entrances. Theoretically, the integration measure shows the cognitive complexity of reaching a space. On the basis of this evidence, space syntax studies argue that integration values can 'predict' the pedestrian use of a space: the easier it is to reach a space, the more often it is going to be used.

Syntactic analysis also shows a direct relation between visibility conditions and integration values (i.e. values that express spaces with the highest levels of physically accessible centrality in the system: the more central, the more reachable, also shown by the highest levels of visual accessibility). Within the external circulation system, the most visible spaces are also the most integrated. They usually appear as main circulation routes and can be designated as the primary circulation system. Those spaces which provide direct access to the buildings define a secondary circulation system which shows low integration values as well as low visibility levels. Because these spaces are linked to the primary circulation system, no area within the campus is left without a potential flow of people according to the natural movement perspective (Hillier et al. 1993).

These findings support the participants' individual experiences, as well as reflecting the implications of an enclosed campus model surrounded by walls, which produces a clear break in the physical and visual continuity of many paths.

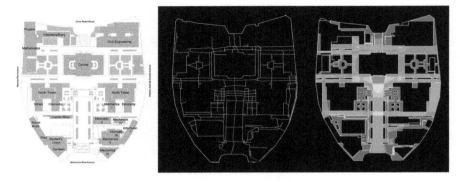

Fig. 1 (From *left* to *right*) The campus plan. Axial map of IST campus. Integration—min: 0.28 (*dark blue*); max: 0.82 (*red*). VGA (visibility graph analysis) connectivity without car parking—min: 8 (*dark blue*); max: 10383 (*red*); grid: 3 m. Connectivity is a syntactic measure that expresses the local accessibility within the space system, i.e. the connections between a space and its immediate surroundings

Among the secondary circulation spaces, stairs appear relatively segregated. At the same time they are not highly visible within the configurational structure. It can be suggested that the visual relations between primary and secondary circulation spaces are rather limited and it is this that results in difficulties of movement within the campus, in particular for those with disabilities.

It is important to mention that in this analysis, stairs and ramps were considered as single and independent lines in terms of movement, once this approach was felt to be more appropriate to the main purpose of the research.

The second level of analysis was based on the exclusion of the stairs and the Campus zones which were accessible exclusively via them. It was assumed that this linear representation and corresponding axial map would represent the true perspective of those travelling in a wheelchair or unable to overcome such barriers. The map (Fig. 2) reveals the importance of the access ramp and shows spaces which are hard to reach, producing a clearly distinct analysis of the Campus.

The visibility graphs in both scenarios (parking zone disregarded or as a barrier) allow one to realise how car parking breaks the links within the Campus. Findings reveal the contrast between the more connected zone, the Campus Alameda—which has no car parking—and the rest of the enclosure which has very low values of connectivity.

On the other hand, by means of the connectivity analysis, it is possible to verify how this spatial system is more cohesive and homogenous when represented without car parking zones. Yellow/orange tones are highlighted and spread all over the zones. By contrast, when the car parking is included, there are predominant low values of connectivity compared to the previous scenarios.

Through axial maps, the Campus network of pathways can be seen from the perspective of those travelling in a wheelchair, revealing the importance of the access ramp. Visibility Graph Analysis proved that car parking harms pedestrian traffic and destroys the natural ability of spaces to allow movement or socialising.

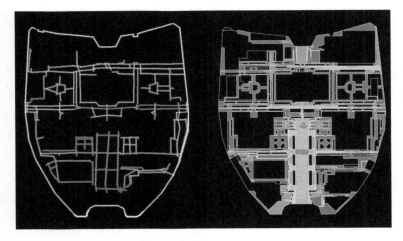

Fig. 2 (From *left* to *right*) axial map of IST campus, excluding stairs. Integration—min: 0.26 (*dark blue*); max: 0.68 (*red*). VGA connectivity—campus with car parking as barriers—min: 1 (*dark blue*); max: 5571 (*red*); grid: 3 m

4.2 Survey of Accessibility Conditions (Walkthroughs)

The survey of accessibility conditions was based on collecting of users' opinions by interviews and walkthroughs, complemented by expert opinions. The interviews conducted with a group of users (including students, teachers and employees with temporary or permanent disability) were intended to identify and characterise each participant and also contain generic questions about accessibility and more specifically about their experience at the IST.

There were also free-form interviews with a group of experts in order to collect different interpretations, perspectives and experiences about accessibility. A metric survey of spatial elements was also developed, intended to list and evaluate stairs, ramps, pathways and sidewalks, and complemented by photographic records, measurements and observations.

The walkthroughs with users were recorded on video and were useful to identify conflicts through a qualitative assessment in which the users classified barriers depending on the effort needed to overcome them: low, medium or intense. The routes were held between the Pavilion and the Canteen, which are the spaces most used by the participants.

The participants in wheelchairs had to take a longer path involving ramps in order to overcome access gaps. Moreover, they had to be helped to climb two steps and to get down the last change of level when the vertical platform elevator did not work properly. The motor handicapped participant took the longest time (37 min 30 s) to go back and forth from the canteen, although it was the third shortest route (628 m). The walk started in front of the Computer Science Department. The participant with a temporary disability identified the bad floor conditions and

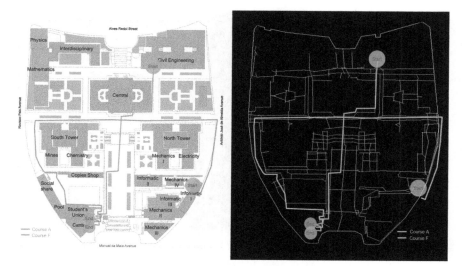

Fig. 3 (*Left*) Routes taken by participants in walkthroughs at IST Campus: walkthroughs 'A' (performed by a visually impaired participant) and 'F' (performed by a wheelchair participant). (*Right*) LoE in Depthmap® (LoE min: 0; LoE max: 15); walkthroughs 'A' and 'F' superimposed to the axial map allows to understand those LoE routes

the fact that the exterior ladders do not have handrails as the major barriers. One of the visually impaired participants once had to be redirected seven times. Not only he did not know the path well, but also he faced several obstacles such as parked cars. The second one revealed that it becomes easier for him to move outside the Campus using the wall as reference.

Based on the video recording of each walkthrough, the route walked by each participant was hand-drawn onto printed plans of the campus. This was used to determine distances as well as the position and the duration of the stops. In addition, behavioural data was related to the syntactic analysis of the campus, based on the external circulation network (Fig. 3).

The circulation network was segmented, as well as the different routes used by each participant. Results have shown that users with disabilities find multiple barriers when crossing the campus: it is not a continuous accessible route.

4.3 Level of Effort

Stairs, ramps and paths were evaluated according to legal requirements and inputs from interviews and walkthroughs. The LoE was evaluated according to intensity, type of floor, the existence of constraints, the existence of railings, projections, uneven floors, slope, dimensional irregularity and presence of obstacles. An

Table 1 Example of the definition and methods for determining LoE

LoE	For intensity, the weight in the final score in the LoE was 35%, with 5 levels corresponding to a scale of 1-9 at incremental values of 2, as presented in the tables.
LoE for ramps	The variable is measured by slope: the first interval was defined on the basis of national legislation for ramps that established a limit of 6% of slope. The remaining intervals are of about 5% and above 20% slope is considered extremely difficult, so it is assigned level 9 (maximum).

Ramp slope and difficulty [%]				
]0;6]	]6; 10]	]10; 15]	]15; 20]	>20
1	3	5	7	9

LoE for stairs	Based on the Law of Blondel: 63 <2 * riser + tread <64. Applying this law, even the easiest stairs had the highest score, so the rule is too strict in some aspects and does not clearly differentiate stairs, or it was badly scaled. Furthermore, stairs with risers or treads too small / large are lumped with stairs with acceptable dimensions. As such, it was necessary to adapt this rule, allowing a wider range of relationships. The step size was also limited. Rules relating to the design of stairs were surveyed to derive a more flexible adaptation of Blondel's Law. Thus, the scale measuring the effort demanded by stairs is: 61 <2 * riser + tread <64 (Neufert 1965), constituting the first interval. The other intervals for a total of 5 categories were defined by an increase of 2 points for every 0.5 above the range given by the rule. Stairs that have risers outside the range [14, 18] cm, or treads less than 28 cm, score one more point to a maximum penalty of 9.

Stair LoE "=2*riser+tread"				
]63; 64[	]62.5; 63] U [64; 64.5[	]62; 62.5] U [64.5 ;65[	]61.5; 62] U [65; 65.5[	]62; 62.5] U [64.5; 65[
1	3	5	7	9

example of the definition and protocols for intensity can be verified in Table 1 (more detailed information is available at Nascimento 2012).

The determination of the evaluation criteria and their weighting in the final value of the LoE demanded the holding of four simulations and a final proposal, obtained by calculations in Excel© and representation in Depthmap© (Nascimento 2012).

It used a table of the survey data, which was completed with the evaluation of the parameters that define the LoE (element by element). This evaluation consisted in assigning points to each item, i.e. according to their geometric characteristics and materials.

The sum scores of each LoE parameter ranged between values of 9 and 26, and subtracting 9, a scale is obtained ranging from 0 to 15 with an increment of 0.5. After simulations, adjusting, adding and removing parameters there was a close approximation perception of LoE users about the space analysed (Nascimento 2012).

There is no element involving a null LoE but then there is no worst value for
LoE (the maximum). Another relevant point is that even on the stairs or ramps
mode (5 or 6, respectively), effort was less than half the maximum of the scale
(15). Since the campus is quite large, most serious situations are abated by sundry
factors.

We also analysed the correlation between the LoE and the barriers identified by
the participants in the walkthroughs. The total value of the barriers (for each step)
was calculated by assigning 1, 2 or 3 points, respectively, to the barriers that
require low, medium or intense effort:

Total value of Barriers = low effort × 1 + medium effort × 2 + intense effort × 3

The linear association between the variables was quantified through the Linear
Correlation Coefficient of Pearson, resulting in a value of $r = 0.5$, which means
that the correlation is Moderate positive. The average correlation between the LoE
and the number of barriers identified by the users may have been influenced by
familiarity or lack of it with the course. There were different routes for users,
according to the different types of disability or impairment. We also have to
consider the fact that the effort is relative for each user (depending on age and
whether they regularly take exercise or not).

5 Conclusion

An inclusive university requires spaces that meet the needs of a wide public,
regardless of age or ability. Any user on a Campus must feel that the space permits
immediate understanding of the different routes and places where they want to go,
without feeling limited in their access to any areas.

The syntactic analysis shows that the spaces on the Campus are not connected
and it is not efficient for the user in terms of orientation and location. VGA proved
that car parking harms pedestrian traffic and destroys the natural ability of spaces
to allow movement or socialising.

Analysis of the routes taken in walkthroughs led to the conclusion that although
there are sites with good accessibility, access to outdoor spaces is not continuous
between the point of origin and the end of a route, which makes autonomy
impossible for the disabled user. Hence we may conclude that the outside spaces of
the Campus have poor accessibility.

LoE calculations completed the survey of some situations not evaluated by
users and allowed the survey to be used as a preliminary assessment of environmental accessibility, without dispensing with subsequent observation in loco and
the perspectives of users. The map which represents the LoE revealed strong
accessibility problems on the Campus, which are the main routes with high values
of LoE.

The use of tools such as space syntax and LoE for accessibility evaluation underlines technical analysis of accessibility problems, and may be developed by collating various media (observation, users' opinion and regulatory requirements). This allows researchers to represent, analyse and simulate how variation in the characteristics of spatial elements influences the accessibility of a space.

These research findings will be presented to the academic administration. The goal is (1) to develop a strategy to implement improvements for physical accessibility within the campus, (2) to monitor the success of those measures and furthermore (3) to achieve, if possible, exemplary procedures applicable to other learning environments. Written reports, a video documentary and a virtual environment are being prepared as well as the ongoing interaction with the academic administration and other decision makers, also including the academic community, making it a matter for future developments.

References

Hanson J (2004) The inclusive city delivering a more accessible urban environment through inclusive design. In: Proceedings of the RICS Cobra 2004 international construction conference: responding to change, York, UK

Hillier B, Hanson J (1984) The social logic of space. Cambridge University Press, Cambridge, UK

Hillier B, Penn A, Hanson J, Grajewski T, Xu J (1993) Natural movement. Environ Plan B 20:29–66

Imrie R, Hall P (2001) Inclusive design: designing and developing accessible environments. Spon Press, London, UK

Levine D (2003) Universal design New York 2. Center for Inclusive Design and Environmental Access, State University at Buffalo, Buffalo, US

Nascimento R (2012) Acessibilidade no espaço construído: do contexto ao utilizador. Caso de estudo: o Campus Alameda do IST. Master Dissertation in Architecture, Instituto Superior Tecnico, Technical University of Lisbon, Portugal

NEP (2012) Factos and números, núcleo de estatística e prospetiva (NEP), Instituto Superior Técnico, Technical University of Lisbon, Portugal

Neufert E (1965) Arte de proyetar en arquitectura. Gustavo Gili, Barcelona, Spain

Olivier M (2008) Understanding disability: from theory to practice. St. Martin's Press, New York, US

Turner A, Doxa M, O'Sullivan D, Penn A (2001) From isovists to visibility graphs: a methodology for the analysis of the architectural space. Environ Plan B: Plan Des 28:103–121

Wagener T, van der Spek S (2006) Accessibility for all: eliminating barriers across Europe. Connected Cities, Rotterdam, The Netherlands

Accessibility in Practice: A Process-Driven Approach to Accessibility

S. Horton and D. Sloan

1 Introduction

Attention to accessibility usually comes into play in the later phases of product development. Accessibility audits are typically performed during quality assurance and user acceptance testing phases. Remediation for any issues identified in the audit usually happens in code. The best fix for many complex accessibility issues may be to revisit the overall design approach, yet reworking designs at this late phase has a significant impact on timelines and processes. Any recommendation involving alternative designs is therefore usually unwelcome. Instead, the issues remain unresolved, or are resolved in a fashion that achieves technical accessibility but offers a compromised user experience.

The best approach to accessible user experience is to integrate accessibility into the design and development process. When accessibility is part of the practice of every member of the product development team, and when accessible features and functionality are built into design, content and code, the result is a product that is accessible and enjoyable for everyone.

As user experience consultants with The Paciello Group (TPG), a US-based consultancy that provides IT accessibility services to companies, agencies and organisations, we see a growing interest in pursuing a more holistic approach to accessibility, one that brings accessibility into the practice of designing and building websites and applications. In this paper, we explore drivers for this interest, including laws and policies that require early attention to accessibility. We present a framework for integrating accessibility efforts into product development processes to bring about a fundamental change in how we approach accessibility.

S. Horton (✉)
Accessible User Experience and Design, The Paciello Group, Nashua, USA
e-mail: shorton@paciellogroup.com

D. Sloan
The Paciello Group, London, UK

P. M. Langdon et al. (eds.), *Inclusive Designing*, DOI: 10.1007/978-3-319-05095-9_10,   105
© Springer International Publishing Switzerland 2014

2 The Value of Accessible User Experience

Good user experience increases satisfaction and enjoyment. Customers who have a good experience place higher value on, and are more loyal to, a product and its producer. Accessible user experience brings the benefits of good user experience to people with disabilities. This, in turn, increases the pool of loyal customers.

2.1 Good UX is Good Business

Good user experience is increasingly recognised as a desirable attribute of digital products, such as websites. It effectively extends the objective components of usability to consider more experiential, subjective qualities that emerge after using a digital product (Hassenzahl 2013). The commercial value of providing a quality user experience is increasingly recognised (Nielsen and Gilutz 2013). Companies that are customer focused, with products that satisfy and delight, see an increase in loyalty behaviours, including repurchase and referrals (Hoisington and Naumann 2003).

From an accessibility standpoint, how can we ensure that people with diverse needs, including people with visual, hearing, mobility and cognitive impairments, have a quality user experience—one that is, in the words of Hassenzahl, *worthwhile*? Put another way, how do we achieve universal usability, with 'more than 90 % of all households as successful users of information and communication services at least once a week'(Shneiderman 2000). Note in Shneiderman's definition, it's not enough to provide products that are merely usable. To reap the shared benefits of accessible user experience, a large and diverse population must have *successful* experiences with digital products.

2.2 Compliance Does Not Ensure Quality

Involving people with disabilities effectively in a digital product development project can lead to valuable and often unexpected insights on how design problems can be solved (Pullin 2010). However, current approaches to Web accessibility tend to be driven by technical guideline conformance (Cooper et al. 2012). A focus on guideline conformance can leave little room to consider context of use, and provides no direct requirement to consider user experience of people with disabilities (Sloan and Kelly 2011). The result is that approaching accessibility from a purely technical conformance perspective can let digital products fall short of providing a quality user experience to people with disabilities (Power et al. 2012).

2.3 Retrofitting is Costly, on All Fronts

Remediating accessibility barriers at the end of the design cycle or after launch is expensive and ineffective. Given time, budget, and technical constraints, the scope for improving the experience of people with disabilities may be limited to reducing the impact of specific discrete accessibility barriers, or worse, creating a parallel version to accommodate specific disabilities. This 'separate, but not equal' approach to achieving compliance produces 'accessibility solutions' that are typically inferior in content and functionality, and are neglected over time, lacking the attention given to corresponding mainstream products (Wentz et al. 2011).

When accessibility efforts are fuelled by compliance needs, prioritised according to perceived impact, and addressed at the end of the product development lifecycle, the quality of the user experience for people with disabilities will have little influence on the work that is done.

2.4 Filling the Gap with Accessible User Experience

Accessible user experience fills the gap between accessibility as a technical quality assurance exercise and accessibility as an attribute of an overall quality customer experience. By adopting a practice of accessible user experience, product designers and developers work to ensure that as many as possible of the target audience, including those with disabilities, can successfully achieve desirable goals when interacting with a product. These goals can be task-oriented, but may also be more experiential. In either case, the outcome will be positive, both for the user and the organisation.

3 Using Standards to Drive Process Change

Some accessibility-related standards contain provisions that are directly related to process. These provide an opportunity to integrate accessibility into culture and processes, as organisations are required to address the needs of people with disabilities throughout the product development lifecycle, and to document their efforts. Fueling these efforts with process-related requirements may help remove attitudinal barriers to integrating accessibility.

For example, the Twenty-First Century Communications and Video Accessibility Act (CVAA) in the United States presents performance objectives that ensure equipment and services are accessible and usable. The rules include the directive that 'Manufacturers and service providers must consider performance objectives … at the design stage as early as possible and must implement such performance objectives, to the extent that they are achievable' (CVAA 2011). Similarly, the

U.S. Information and Communication Technology Standards and Guidelines, a.k.a. Section 508 Refresh, requires that product developers include accessibility in design and development. These standards also require that people with disabilities are included in the product design and evaluation process, and that the needs of people with disabilities are considered in market research efforts (Section 508 Refresh 2011). Complementary to technical standards for measuring accessibility of digital resources, other standards specifically focus on the process of accessible technology design, encouraging inclusive design throughout the system design lifecycle. As an example of a process standard, BS 8878 was launched in the UK in 2010, to define a procedure for commissioning and implementing accessible websites—whether developed internally by an organisation, or outsourced (British Standards Institute 2010). Thus, BS 8878 is written to support managerial staff throughout the project lifecycle, and does not demand in-depth technical knowledge.

We see these process-related standards, guidelines and reporting requirements as providing an opportunity for organisations to adopt holistic support for accessible user experience, and build in accessibility from the start. Given a requirement to consult with people with disabilities, include people with disabilities in user research, and measure accessibility and usability of products against performance objectives—these are tasks that cannot be undertaken and reported on after the fact without slowing timelines and expending significant additional resources. When attention to the user experience of people with disabilities is part of the overall process, requirements will be easier to achieve, reporting will be more credible, and the outcome will be a better process and product, for everyone.

4 A Practice of Accessible User Experience

Designing and engineering a quality user experience for people with disabilities requires an organisational commitment, shared responsibility and accountability for accessibility, and supporting resources to establish a practice of accessible user experience (Horton and Quesenbery 2014).

A key challenge to establishing a practice of accessibility within an organisation is overcoming perceptions about the difficulty, complexity, effort, and cost of achieving accessibility. The challenge of changing current practices to the ones that support accessibility is often seen as insurmountable.

To illustrate this challenge we provide the following scenario. A large, decentralised organisation receives accessibility audits of its websites. Initially, the audits are seen as too general, prompting feedback that site developers require specific code changes. Subsequent audits include detailed feedback, identifying exact locations of problems and code fixes. The detailed audits raise other concerns—namely, that there is too much that needs fixing. When asked what approach would work, the response is that nothing will solve this—the culture is too complex, with too many devolved responsibilities and competing objectives, to apply an effective organisation-wide strategy for achieving accessibility.

This scenario illustrates the negative perceptions that arise from approaching accessibility as an audit and remediation process—one that involves identifying problems and fixing them. It also recognises the challenges of implementing large-scale process change within an organisation. However, process change is what makes accessibility an achievable objective. In the section on partnering with accessibility experts, we discuss the role accessibility trainers and integrators—consultants who are sympathetic to organisational culture, and through practical and pragmatic approaches position accessibility as an achievable objective. In the next section we talk about the role of organisations in making a strategic commitment to accessibility, using a framework of disruptive innovation to effect process change.

4.1 Make an Organisational Commitment to Accessibility

For most companies, introducing accessibility into the product development process may be experienced as a disruptive innovation, as it introduces a new value proposition to the organisational culture. Initiatives that fail to appreciate and work with disruptive attributes may not be successful (Christenson 2011). Therefore, it is crucial that organisational leadership recognises the commitment as disruptive, and follows innovation adoption best practices when initiating and implementing accessibility.

Everett Rodgers offers a model for introducing an innovation process within an organisation (Rogers 2003). Here, we use his model as the basis for implementing accessibility within an organisation. We show the Rodgers phases and definitions, mapped to the context of adopting a practice of accessible user experience (shown in italics).

In the Initiation phases, the organisation acknowledges that current design and development processes do not produce accessible products:

- Agenda-setting: General organisational problems that may create a perceived need for innovation—*Current design and development practices do not produce accessible products.*
- Matching: Fitting the problem from the organisation's agenda with an innovation—*Adopting a practice of accessible user experience.*

In the Implementation phases, the organisation revisits current organisational structures and design and development practices, and creates a plan for successfully adopting and integrating accessibility into practice:

- Redefining/Restructuring: The innovation is modified and reinvented to fit the organisation, and organisational structures are altered—*Accessible user experience is defined within the context of the organisation and the digital technologies it produces or uses.*

- Clarifying: The relationship between the organisation and the innovation is defined more clearly—*Responsibilities and required accessibility skill sets and tools for delivering objectives are defined and assigned within the product team.*
- Routinising: The innovation becomes an ongoing element in the organisation's activities, and loses its identity—*The organisation builds products with accessibility built-in; accessibility is an integral part of the quality assurance process.*

4.2 Establish Roles and Responsibilities

Product development teams come in many shapes and sizes. Product development processes range from distinct teams responsible for discrete aspects of a product to integrated teams that share responsibility for multiple dimensions. Some 'teams' are composed of one person who manages all aspects of product development. Regardless, for accessible user experience, the responsibilities incumbent on each role must be defined and assimilated by the responsible party.

A systematic, guidelines-based approach is to map the Web Accessibility Initiatives Web Content Accessibility Guidelines (WCAG 2.0) (W3C World Wide Web Consortium 2008) success criteria to the roles on the product team. The WAI-Engage Community Group is currently exploring this approach on their collaborative Wiki. In the accessibility responsibility breakdown (ARB) matrices, responsibility for meeting the WCAG 2.0 levels and success criteria are mapped to different project management, research, design, content, and development roles. This could be a useful starting point for establishing primary roles and responsibilities within a project team; although care should be taken to treat accessibility as a holistic and shared process across the project team and beyond.

4.3 Establish an Accessibility Infrastructure

Take steps to integrate accessibility into the infrastructure of the organisation. Like whiteboards and post-its, OmniGraffle and Photoshop, GitHub and SublimeText, make accessibility resources part of the product development toolkit.

- Organisational policy: Every organisation has policies—whether explicit operational policies, or implicit practices that are part of the organisational culture. Standards-based development is a good example: many product developers hold developing to standards as policy. An organisational policy can go far in articulating an organisational commitment on accessibility, and building an 'accessibility-first' mindset across the organisation (Kline 2011; Feingold 2013).

- Content strategy: Organisations have a growing appreciation for the importance of content strategy. Including accessibility in an overall content strategy makes accessibility a strategic priority (Kissane 2010).
- Code repositories: Managing an accessible code repository for common user interface components, such as menus, tabs, date pickers, is a worthwhile investment. It streamlines development time, and provides a better, more consistent, more accessible user experience. There is a range of publicly available examples of accessible user interface design patterns, such as the Accessible jQuery User Interface Components Demonstration (http://access.aol.com/aegis/), which can be incorporated into internal repositories.
- Style guides: Rules and conventions that support best practices help teams produce better products. Integrating accessibility into style guides helps build in accessibility features by following established and shared best practices (Lynch and Horton 2009).
- Content and development tools: Tools can make accessibility a struggle, or something that is baked in. Look for content management software that supports accessible templates and promotes accessible content production: for example, adding equivalent text to images and marking up content using semantic markup. Look for development tools that support standards-based markup and code validation.

4.4 Build Accessibility Expertise on the Product Development Team

Ultimately, accessible user experience requires a commitment from every member of the product development team, and everyone shares accountability for providing accessibility within his or her realm of responsibility. When recruiting new team members, include experience with accessibility as a job requirement, and list accessibility-related tasks in the list of job responsibilities.

Assess familiarity and proficiency with accessibility at hiring time—whether for UX practitioners, developers or managers. Has the applicant been exposed to accessibility training and education? How much experience does she have with working on projects involving accessibility? Prospective employees should be able to present direct evidence of involvement with the knowledge of accessibility practices. Many programmes and positions include accessibility in the description, but in reality provide minimal exposure.

The scarcity of any formal accessibility qualifications that could be used to assess an individual's professional skills in accessibility is a hindrance to assessing prospective candidates. A professional IT Accessibility society has been in discussion, although arguments against this approach focus on the need to promote accessibility as a core professional skill rather than a specialty (Rush 2012).

4.5 Involve People with Disabilities in the Product Development Process

Strategic product development begins with a definition of purpose, goals and target audience. This important step creates a roadmap for the product development lifecycle, and a framework for decision-making. For accessible user experience, it is critical that people with disabilities are considered among the target audience. When the user experience of people with disabilities is integral to the strategic goals of the product, other pieces of the accessible user experience process fall into place. While participation of people with disabilities in usability evaluations towards the end of the design process can help to validate accessibility-related decisions taken earlier on, their involvement at an earlier stage can potentially lead to valuable insights and innovative approaches to design problems.

For instance, user research helps teams identify content, functionality, and design strategies that will be welcomed by and useful to the product's target audience. Including people with disabilities in user research helps ensure that accessibility is integrated into all aspects of a product. The needs of people with disabilities are not just accommodated, but rather directly addressed, in decisions about content, design and functionality.

Involving people with disabilities does not mean consulting individually with people representing each and every type of disability, any more than user research means surveying a group large enough to fully represent the entire target audience. The range of needs and preferences for people with disabilities is the same as with all people—vast, diverse and individualised—and we can learn a great deal and make informed design decisions by bringing in perspectives from one person or one hundred. Whether through interviews, focus groups, surveys or personas, for accessible user experience the only difference is that the definition of users includes people with disabilities.

4.6 Establish Partnerships with Accessibility Experts

Accessibility consultants are often brought into a project in an audit capacity, looking for and reporting on compliance violations. Cast in the role of 'accessibility cops', consultants often have limited positive impact on organisational accessibility awareness and processes. A better approach is to partner with accessibility experts. Bring consultants into projects early, as accessibility partners rather than compliance officers. Have accessibility expertise on projects, alongside strategy, design, programming and content experts.

An informed accessibility perspective is also important to help guide any process and method changes that may be needed to integrate accessibility into practice. For instance, user-centred research and design methods may need to be modified in order to recognise constraints related to the nature of participants' impairments or other characteristics, and how that impacts on data collection mechanisms and the type of data that can be collected. Prior (2011) discusses how methods can be adapted to effectively involve people with severe speech and physical impairments as co-designers; Dickinson et al. (2007) provide valuable advice on involving older people in user-centred design activities.

For an effective partnership, accessibility experts should be mindful of the challenges facing designers and developers, particularly those relating to business-related constraints on design requirements, budgets and timescales. Rather than point out all the ways a product fails to meet accessibility standards, give useful, actionable and achievable guidance on ways to capitalise on opportunities, such as new ways to involve a customer base of people with disabilities earlier in the design and development process, and to remedy issues. Look for opportunities to inform and train designers and developers on accessible user experience patterns and methods. Bring accessibility to the table as a challenge that is intriguing, worthwhile—and solvable.

5 Conclusions

People with disabilities are customers with demands and expectations for quality experiences like anyone else, or potential customers yet to be engaged. Providing an enjoyable and successful experience of products to customers has a high return on investment. Providing a compromised experience to people with disabilities is a lost opportunity, as well as a potential compliance issue. The most beneficial way to address accessibility is by broadening the definition of 'target audience' to encompass the broad and ever-shifting range of ability, having each member of the product develop team assume responsibility and accountability for accessibility features, and providing resources and expert guidance to support a practice of accessible user experience.

Our accessible user experience framework is intended to guide organisations in planning and implementing a strategy for integrating accessibility throughout the design lifecycle, and in particular, ensuring that user experience design and inclusive design activities can be combined for positive benefit. Future work, as part of the authors' roles as accessible user experience consultants with TPG, will involve supporting adoption of the framework and measuring its effectiveness. Key success criteria will include the quality of accessible user experience of digital products developed using the framework and the extent to which the framework can be adopted in practice by organisations in different sectors. As an organisation offering accessibility expertise, we will also explore the most effective and productive ways of working with organisations to implement the framework element, 'establish partnerships with accessibility experts'.

References

British Standards Institute (2010) BS 8878:2010—web accessibility code of practice. BSI, London

Christenson C (2011) The innovator's dilemma: the revolutionary book that will change the way you do business. HarperBusiness, New York

Cooper M, Sloan D, Kelly B, Lewthwaite S (2012) A challenge to web accessibility metrics and guidelines: putting people and processes first. In: Proceedings of the 2012 International Cross-Disciplinary Conference on Web Accessibility (W4A), Lyon, France

CVAA (2011) Final report and order as released Oct 7, 2011. Available at: www.fcc.gov/document/accessibility-rules-advanced-communications-services-0. Accessed 29 Oct 2013

Dickinson A, Arnott J, Prior S (2007) Methods for human-computer interaction research with older people. Behav Inf Technol 26(4):343–352

Feingolf L (2013) Accessibility information pages show commitment to all site users. Available at: http://lflegal.com/2013/02/access-info-pages/. Accessed 29 Oct 2013

Hassenzahl M (2013) User experience and experience design. In: Soegaard M, Dam R (eds) The encyclopedia of human-computer interaction (2nd ed) The Interaction Design Foundation, Aarhus, Denmark. Available at: www.interaction-design.org/encyclopedia/user_experience_and_experience_design.html. Accessed 29 Oct 2013

Hoisington S, Neumann E (2003) The loyalty elephant. Qual Prog 36(2):33–41

Horton S, Quesenbery W (2014) A web for everyone: designing accessible user experiences. Rosenfeld Media, Brooklyn

Kissane E (2010) The elements of content strategy. A Book Apart, New York

Kline J (2011) Strategic accessibility: enabling the organization. Live Oak Book Company, Palo Alto

Lynch P, Horton S (2009) Web style guide: basic design principles for creating web sites. Yale University Press, New Haven. http://www.webstyleguide.com

Nielsen J, Gilutz S: Return on investment (ROI) for usability. Available at: http://www.nngroup.com/reports/usability-return-on-investment-roi/. Accessed 29 Oct 2013

Power C, Freire A, Petrie H, Swallow D (2012) Guidelines are only a half of the story: accessibility problems encountered by blind users on the web. In: Proceedings of the SIGCHI conference on human factors in computing systems, Austin

Prior S (2011) Towards the full inclusion of people with severe speech and physical impairments in the design of augmentative and alternative communication software. PhD Thesis, University of Dundee, UK

Rogers E (2003) Diffusion of innovations. Free Press, New York

Rush S (2012) Considering the case for creating an international society of accessibility professionals. Knowbility blog. Available at: http://blog.knowbility.org/content/considering-the-case-for-creating-an-international-society-of-accessibility-professionals/. Accessed 29 Oct 2013

Section 508 Refresh (2011) Information and communication technology (ICT) standards and guidelines. Available at: www.access-board.gov/guidelines-and-standards/communications-and-it/about-the-ict-refresh/draft-rule-2011. Accessed 29 Oct 2013

Shneiderman B (2000) Universal usability. Commun ACM 43(5):84–91

Sloan D, Kelly B (2011) Web accessibility metrics for a post digital world. In: Proceedings of the W3C web accessibility initiative research and development working group website accessibility metrics online symposium. Available at: http://opus.bath.ac.uk/27541/. Accessed 29 Oct 2013

WAI-Engage Community Group Accessibility responsibility breakdown. Available at: www.w3.org/community/wai-engage/wiki/Accessibility_Responsibility_Breakdown. Accessed 29 Oct 2013

W3C World Wide Web Consortium (2008) Web content accessibility guidelines 2.0 (WCAG 2.0). Available at: www.w3.org/TR/WCAG20/. Accessed 29 Oct 2013

Wentz B, Jaeger P, Lazar J (2011) Retrofitting accessibility: the legal inequality of after–the–fact online access for persons with disabilities in the United States. First Monday 16(11). Available at: http://firstmonday.org/ojs/index.php/fm/article/view/3666/3077. Accessed 29 Oct 2013)

Part IV
Designing Inclusive Spaces

Improving Residents' Satisfaction in Care Homes: What to Prioritise?

S. A. Shamshirsaz and H. Dong

Abstract Information on consumer satisfaction is essential for quality improvement in care homes. We identified 12 essential needs based on our interviews with 15 care home residents in the UK, and conduced a Kano questionnaire survey of 102 residents from 35 care homes in the London region. Three 'delighters' from the 12 essential needs were further identified, i.e. accessible equipment, home-like environment, and suitable design. It is believed that the improvement of any of these three factors will significantly increase residents' satisfaction.

1 Introduction

Information on consumer satisfaction, whether the consumers are the residents or the family members, is essential for quality improvement in care homes (Mansfield et al. 2000; Castle 2007). The most effective way to improve quality in an organisation is understanding customer needs and expectations and trying to meet or exceed them (Deming 1986).

Over half of the health care beds in the United Kingdom belong to independent care homes for older people. A substantial corporate care home sector has emerged over the last two decades, contributing 57 % of private sector capacity for older and physically disabled people (Laing 2012). When an elderly person considers moving to a care home, a test will be carried out by the local authority to assess his/her level

S. A. Shamshirsaz
Inclusive Design Research Group, School of Engineering and Design,
Brunel University, London, UK
e-mail: sanaz.shamshirsaz@gmail.com

H. Dong (✉)
Inclusive Design Research Centre, College of Design and Innovation,
Tongji University, Shanghai, China
e-mail: donghua@tongji.edu.cn

P. M. Langdon et al. (eds.), *Inclusive Designing*, DOI: 10.1007/978-3-319-05095-9_11,   119
© Springer International Publishing Switzerland 2014

Table 1 Highly ranked demanded qualities	No.	Demanded qualities
	1	Empathic staff
	2	Social interaction
	3	Autonomy
	4	Accessible equipment
	5	Meals
	6	Safety
	7	Family support
	8	Accurate medical care
	9	Involvement
	10	Home-like environment
	11	Daily living activities
	12	Suitable design

of needs, and his/her income and assets are taken into account for the payment of care. This study was conducted with independent care homes in the UK.

Quality function deployment (QFD) is a set of tools that are combined to form a process for obtaining and integrating the voices of customers (VoC) into every feature of the design and conveyance of goods and services. Using VoC, the lead author interviewed 15 residents in three different care homes in the UK, and identified 28 primary needs and requirements, i.e. Demanded Qualities (DQs), of which 12 (Table 1) were further determined to be the most significant ones (Shamshirsaz et al. 2012).

The aim of this study was to examine the most important needs (i.e. demanded qualities) with a larger group of residents in order to establish priorities for improving residents' satisfaction in care homes.

1.1 Residents' Satisfaction

The importance of consumer satisfaction is clearly growing in all areas of health care. As Mansfield et al. (2000) noted, 'Satisfaction surveys present a potent tool to improve long-term care services'. Resident satisfaction surveys can shed light on areas that need improvement and identify areas or care processes that consumers find lacking. The results of resident satisfaction surveys can influence the sensitivity of nursing home staff to human needs and residents' quality of life (Howard et al. 2001). For enhancing customer satisfaction, the aim should be not only to meet customer requirements but also to recognise needs (Bilgili et al. 2011).

However, capturing residents' satisfaction presents challenges. First, most residents in care homes are frail and they may suffer from cognitive impairment. Some researchers believe that residents who suffer from cognitive impairment would experience difficulty answering questions reliably (Davis et al. 1997). Therefore, a large number of studies have focused on families and staff rather than residents. This tendency of using proxy users instead of elderly people themselves can also be found in other areas. Second, because many residents see themselves as vulnerable to retaliation from staff, they may be reluctant to respond truthfully. Researchers find that care home residents are less likely than other kinds of consumers to express dissatisfaction (Castle and Engberg 2004).

Numerous studies suggest that family members and staff have different concerns and perspectives related to quality of care and services received by residents. So they cannot be direct proxies for the residents' experiences because their ratings on salient features of a long-term care facility differ from those of the consumers directly affected by such features (Gasquet et al. 2003; Ejaz et al. 2003; Lowe et al. 2003). Albeit the viewpoints of family and staff are important, 'consumer-centred care' requires a focus on the viewpoints of residents, who may disagree with the viewpoints of staff, family and others about the importance of many features (Peak and Sinclair 2002).

1.2 Kano Model

Customers' feedback is critical because they are the focal points for quality control and improvement (Mitra 2008). In a competitive market environment, providing a high quality of service has been identified as a fundamental factor for achieving customer satisfaction (Carnevalli et al. 2010).

The relationship between service attributes, performance and customer satisfaction was determined as linear and one-dimensional (Metzler and Hinterhuber 1998). It was assumed that the less customers noticed a service or a product less satisfied they would be and vice versa. However, meeting a customer's expectations, even at a very high level, does not always result automatically in his or her satisfaction. Other kinds of expectations that define the perceived attributes of the service or product are also linked to customer satisfaction. Kano et al. (1984) identified that in terms of overall customer satisfaction and dissatisfaction. The two concepts are independent, and service attributes do not contribute equally to them. With regard to customer satisfaction or dissatisfaction, the Kano model (Fig. 1) rejects a simple linear representation of the relationship. Instead, it distinguishes three levels of customer needs, i.e.:

1. basic needs ('dissatisfiers'),
2. expected or one-dimensional needs ('satisfiers'), and
3. excitement needs (or 'delighters').

Basic needs are so obvious to the customer that he or she may feel they do not warrant a mention (hence 'unspoken')—until or unless the service or product neglects to deliver them. Meeting these needs may not steeply increase customer satisfaction; however, failing to do so will undoubtedly provoke a strong negative customer reaction.

Expected or one-dimensional needs are related to customers' conscious expectations, and the product's or service's attributes have a linear relationship

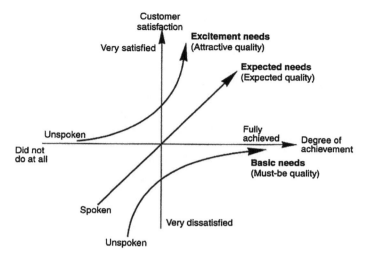

Fig. 1 Kano model (*Source* Kvist and Klefsjo (2006)—based on the original model presented by Kano et al. 1984)

with overall satisfaction; that is to say, the more effectively the product or service meets the customer's needs, the higher his or her level of satisfaction will be.

Excitement needs refer to those needs that can surprise and delight customers. This type of quality is also described as the 'wow' factor (Shahin et al. 2012). The customer is not aware of these unexpected needs (hence 'unspoken'), and therefore their absence does not cause dissatisfaction. However, customer satisfaction will be increased when these attributes are provided. Service providers who realise and satisfy these "excitement needs" create what is described as "attractive quality". This category of needs is not strictly limited to innovations in technical products or services; it includes well-delivered or executed services that might be seen as exciting (Kvist and Klefsjo 2006).

Customer needs can be categorised through the use of the Kano questionnaire. In the Kano questionnaire, each question has two parts (a pair), the first part asks the customer how he/she feels if the feature of product or service is provided; and the second part ask how the customer feels if the feature is not provided. An example of the Kano questionnaire is shown in Table 2.

Customers' responses to paired Kano questionnaires fall into three levels: basic (B), one-dimensional (O) and excitement requirements (E). The Kano evaluation table (Table 3) is used to interpret the result of the Kano questionnaire. Based on the frequency of answers, the customer needs can be classified (Kano et al. 1984): i.e. B (basic), O (one-dimensional), E (excitement); and one-dimensional needs also include indifferent attributes (I) of a product or service which have no significant impact on customer satisfaction or dissatisfaction, and reverse requirements (R) which indicate areas where the customer is less satisfied due to the presence of the product or service attributes and vice versa.

Table 2 Paired questions for Kano questionnaire

Kano questionnaire		
1a	How would you feel if X was provided?	5. I like it
		4. I expect it
		3. I feel neutral
		2. I can tolerate it
		1. I dislike it
1b	How would you feel if X was not provided?	5. I like it
		4. I expect it
		3. I feel neutral
		2. I can tolerate it
		1. I dislike it

Table 3 Kano evaluation table

Interpretation of Kano questions	How do you feel if X was not provided?				
	5. I like it	4. I expect it	3. I feel neutral	2. I can tolerate it	1. I dislike it
How do you feel if X was provided?					
5. I like it		E	E	E	O
4. I expect it	R		I	I	B
3. I feel neutral	R	I		I	B
2. I can tolerate it	R	I	I		B
1. I dislike it	R	R	R	R	

E exciting, *B* basic, *O* one-dimensional, *I* indifferent, *R* reverse

2 Methods

This study examines care homes' quality attributes by using a questionnaire survey. The Kano model was adopted to gather residents' perceptions and classify quality attributes.

2.1 Sampling

In total, 250 care homes within or near the London region were selected randomly for the distribution of the questionnaires. The target was to allow the Ph.D. researcher to visit the care homes when necessary. The questionnaires were accepted by 59 care homes. Residents from 35 care homes actually participated in the survey. The selection of the participants was according to the following criteria: (1) an interest in participating in the research, (2) good understanding of English, (3) having no Alzheimer's or other known cognitive impairment, (4) having enough energy to participate in the study and (5) having resided in the care home for at least 4 weeks.

Respondents completed questionnaires voluntarily (some were visited by the Ph.D. researcher to encourage participation) within 7 months (February to August 2012). The total number of respondents was 102, coming from care home ranging from 10 beds to the largest, with 160 beds.

2.2 Designing the Questionnaires

The data captured from residents by the authors previously (Shamshirsaz et al. 2012) were qualitative in nature. The 12 highly ranked demanded qualities (DQs— see Table 1) were used to design the survey questionnaire which has three parts:

Part 1 was designed to identify the relative importance of each DQ to the others. It gave residents the opportunity to choose one statement for each DQ, using a 5-point Likert scale (1: does not matter; 5: matters very strongly).

Part 2 was to evaluate the level of residents' satisfaction with given services in the care home as well as its competitor (in this case, another care home in the same region, with similar service provisions) on each of the selected DQs. It adopted a five-point scale (1 for poor; 5 for excellent).

Part 3 contains the Kano questionnaire. The modified triple-choice questions (Table 4) were used instead of five choices to make it easier for residents to answer (Aghlmand et al. 2010).

As the modified Kano questionnaire was used, there are nine possible combinations for the paired questions (Table 5).

3 Analysis and Results

The medians of the answers to Part 1 and Part 2 of the questionnaires were calculated. The median is utilised, as the questionnaire results were discrete ordinal data (Likert scale) where medians are more suitable than means (Aghlmand et al. 2010).

The results of Part 1 show that four demanded qualities, i.e. empathic staff, social interaction, safety and home-like environment, were rated as the most important requirements. Involvement was considered as the least important.

The results of Part 2 demonstrate that the services related to six demanded qualities were ranked better in the chosen care home than its competitors, and these are empathic staff, autonomy, meals, family support, accurate medical care, home-like environment.

Part 3, the Kano questionnaire, was analysed by counting frequencies. Within the 12 demanded qualities, three were identified as the excitement attributes, i.e.

- accessible equipment,
- home-like environment, and
- suitable design.

Table 4 Modified Kano questionnaire

Kano questionnaire		
1a	How would you feel if X was provided?	3. I like it
		2. I expect it
		1. I feel neutral
1b	How would you feel if X was not provided?	3. I feel neutral
		2. I can tolerate it
		1. I dislike it

Table 5 Modified Kano evaluation table

Interpretation of Kano questions	How would you feel if X was not provided?		
	3. I feel neutral	2. I can tolerate it	1. I dislike it
How would you feel if X was provided?			
3. I like it	E	E	O
2. I expect it	I	I	B
1. I feel neutral	I	I	B

E exciting, *B* basic, *O* one-dimensional, *I* indifferent

Six were one-dimensional requirement, including empathic staff, social interaction, autonomy, daily living activities and family support, and one indifferent attribute, i.e. involvement. Basic requirements include meals, safety and accurate medical care.

4 Prioritising Needs for a Specific Care Home

Quality improvement for a care home will need to take into consideration the care home's specific situation (e.g. current service provision, competitors, and quality target). A quality planning table (QPT—see Table 6) can be utilised to convert the voice of the customer into the voice of the design team or the organisation (Chaplin 2000).

In this section, a specific care home was selected as an example to show how the survey data can be used in the quality planning table to prioritise items for improvements. The care home is located in Peterborough, with 11 beds; and 11 residents answered the questionnaire. Another care home, also in Peterborough, with 22 beds, was regarded as its comparison. In the 22-bed care home, 11 residents answered the questionnaire.

In Table 6, the left four columns include the data captured from the residents' survey, i.e. A. demanded qualities (Part 1 data—medians from all respondents), B. importance ratings (Part 2 data, medians from the 11 participants from the 11-bed care home), C. competitive ranking (Part 2 data, medians from the 11 participants

Table 6 Quality planning table for a specific care home

Quality planning table	A. Customer importance rating	B. Competitor rating	C. Our performance rating	D. DQ Kano level	E. Target	F. Improvement ratio	G. Sales point	H. DQ absolute importance	I. DQ relative importance	J. DQ relative importance (%)
1. Empathic staff	5.0	3.5	4.0	O	5.0	1.25	1.5	9.4	0.13	11
2. Social interaction	5.0	3.0	3.0	O	4.5	1.50	1.2	9.0	0.12	11
3. Autonomy	4.0	3.0	3.5	O	5.0	1.43	1.2	6.9	0.09	8
4. Accessible equipment	4.0	3.0	3.0	E	4.0	1.33	1.2	6.4	0.09	8
5. Meals	4.5	3.5	5.0	B	5.0	1.00	1.2	5.4	0.07	6
6. Safety	5.0	4.0	4.0	B	5.0	1.25	1.5	9.4	0.13	11
7. Family support	4.0	4.0	5.0	O	5.0	1.00	1.2	4.8	0.07	6
8. Accurate medical care	4.0	3.5	4.0	B	5.0	1.25	1.2	6.0	0.08	7
9. Involvement	3.0	2.0	2.0	I	3.0	1.50	1.0	4.5	0.06	5
10. Home-like environment	5.0	3.0	3.5	E	4.5	1.29	1.2	7.7	0.11	9
11. Daily living activities	4.0	3.0	3.0	O	4.0	1.33	1.2	6.4	0.09	8
12. Suitable design	4.0	3.0	3.0	E	5.0	1.67	1.2	8.0	0.11	10

from the 22-bed care home) and D. Kano levels of demanded qualities (Part 3 data based on all respondents).

The fifth to the seventh columns of the quality planning table contain specific information discussed with, and agreed by, the care home manager. They include:

- E. target for improvement: Based on residents' data (shown in the left side of the QPT), a target for residents' satisfaction is set for each demanded quality.
- F. ratio of improvement: Demonstrates the potential improvement for each demanded quality, calculated by dividing the target value by the customer perception obtained from the survey.
- G. sales points: Focusing on all demanded qualities may require heavy resources; sales points are utilised to determine DQs that have a potential to improve. Sales points are rated as 1.5 (strong), 1.2 (medium) and 1.0 (weak) to amplify the weighting of the demanded quality to distinguish a product/service from the competition (Tontini 2007).

According to the quality planning table (Table 6), empathic staff, social interaction and safety are identified as the most important DQs for the selected care home, while involvement is the least important DQ for it. When resources are limited, the target of resident satisfaction should therefore be provided by attentiveness to the three highly ranked DQs.

5 Discussion and Conclusions

Assessment of quality can become more reliable and legitimate by integrating residents' views, experiences and perceptions as well as professional and other stakeholder opinions. This study proves the effectiveness of assessing care home services through asking the consumers about their satisfaction with the services provided. It is crucial for care homes to focus resources on the most important needs of residents.

In published survey studies of residents and families, different dimensions of requirements are theoretically identified. However, no study has prioritised residents' expressed needs or assessed the level of requirements from the perspective of residents in care homes. This study has utilised the quality function deployment process and tools, in combination with the Kano model, to identify the relative importance of the identified requirements (i.e. demanded qualities).

Based on the data collected from 102 residents, three delighters were identified, i.e. accessible equipment, home-like environment, and suitable design. It is believed that the improvement of any of these three factors will significantly increase residents' satisfaction.

The quality planning table tool demonstrates how to use the survey data in combination with specific quality targets to prioritise items for improvement.

Although the data were collected from a relatively small number of care homes (qualitative data from three care homes, and quantitative data from 35 care homes), which are not geographically representative, the main residents' requirements identified are similar to published studies. These requirements were further differentiated and rated, and such information can assist care homes to set targets for improving residents' satisfaction. The methodology can also be applied to different care homes for quality planning.

References

Aghlmand S, Lameei A, Small R (2010) A hands-on experience of the voice of customer analysis in maternity care from Iran. Int J Health Care Qual Assur 23:153–170

Bilgilia B, Erciú A, Üna S (2011) Kano model application in new product development and customer satisfaction (adaptation of traditional art of tile making to jewelries). Procedia Soc Behav Sci 24:829–846

Carnevalli JA, Miguel PAC, Calarge FAJ (2010) Axiomatic design application for minimising the difficulties of QFD usage. Int J Prod Econ 125:1–12

Castle NG (2007) A review of satisfaction instruments used in long-term care settings. J Aging Soc Policy 19:9–41

Castle NG, Engberg J (2004) Response formats and satisfaction surveys for elders. Gerontol Soc Am 44:358–367

Chaplin E, Terninko J (2000) Customer driven healthcare: QFD for process improvement and cost reduction. ASQ Quality Press, Milwaukee

Davis M, Sebastian J, Tschetter J (1997) Measuring quality of nursing home service: residents' perspective. Psychol Rep 81:531–542

Deming WE (1986) Out of the crisis. MIT Press, Cambridge

Ejaz F, Straker J, Fox K, Swami S (2003) Developing a satisfaction survey for families of Ohio's nursing home residents. Gerontologist 43:447–458

Gasquet I, Dehe S, Gaudebout P, Falissard B (2003) Regular visitors are not good substitutes for assessment of elderly patient satisfaction with nursing home care and services. J Gerontol: Biol Sci Med Sci 58:1036–1041

Howard PB, Clark JJ, Rayens MK, Hines-Martin V, Weaver P et al (2001) Consumer satisfaction with services in a regional psychiatric hospital: a collaborative research project in Kentucky. Arch Psychiatr Nurs 15:10–23

Kano N, Seraku N, Takahashi F, Tsuji S (1984) Attractive quality and must-be quality, English translation of the article: Miryokuteki Hinshitu to Atarima Hinshitu. Hinshitu. J J Soc Qual Control 14:39–48

Kvist AJ, Klefsjo B (2006) Which service quality dimensions are important in inbound tourism? a case study in a peripheral location. Manag Serv Qual 16:520–537

Laing W (2012) Care of elderly people: UK Market Survey 2011/12, 24th edn. Laing and Buisson, London

Lowe TJ, Lucas JA, Castle NG, Robinson JP, Crystal S (2003) Consumer satisfaction in long-term care: state initiatives in nursing homes and assisted living facilities. Gerontol Soc Am 43:883–896

Matzler K, Hinterhuber H (1998) How to make product development project more successful by integrating Kano's model of customer satisfaction into quality function deployment. Technovation 18:25–38

Mansfield JC, Ejaz FK, Werner P (2000) Satisfaction surveys in long-term care. Springer, New York

Mitra A (2008) Fundamentals of quality control and improvement, 3rd edn. Wiley, Hoboken

Peak T, Sinclair SV (2002) Using customer satisfaction surveys to improve quality of care in nursing homes. Health Soc Work 27:75–79

Shahin A, Pourhamidi M, Antony J, Hyun Park S (2012) Typology of Kano models: a critical review of literature and proposition of a revised model. Int J Qual Reliab Manag 30:341–358

Shamshirsaz SA, Dong H, Aghlmand S (2012) Using voice of customer (VoC) techniques to explore residents' satisfaction in care homes in the UK. In: Proceedings of the 4th international conference for Universal Design, Fukuoka, Japan

Tontini G (2007) Integrating the Kano model and QFD for designing new products. Total Qual Manag Bus Excell 18:599–612

Being Transported into the Unknown: How Patients Experience the Route to the Operation Room

M. Annemans, C. H. Van Audenhove, H. Vermolen and A. Heylighen

Abstract When you are admitted to a hospital, you enter an often-unfamiliar environment. From a person, you suddenly become a patient. Especially when being transported in a bed, you lose control over where you go and how you interact with the space around you. Many people feel anxious and vulnerable under these conditions. Often the built environment adds to this confusion, whereas it should support patients during their hospital experience. To explore the relation between motion and space, we present a concise literature review on mobile and spatial experience, and sketch the theoretical background on researching motion. To actually study the impact of space and spatial elements on patients' mobile hospital experience, we interviewed 12 patients admitted to the day ward of a general city hospital. The interviews were supported by videos of the transport, which the researchers had previously made. Studying the role of motion in patients' spatial experience of the hospital appeared not to be an easy task. Both methodologically and content wise, we encountered some challenges in bringing to the surface and unravelling those experiences linked to spatial aspects. Despite what one could expect, being moved in a bed does not reduce but rather expands patients' sensory perception. An important interrelation between space, time and people, especially during the transport, was found. By designing space architects thus may have a strong influence on time perception and people interactions.

M. Annemans (✉) · A. Heylighen
Department of Architecture, Research[x]Design, University of Leuven (KU Leuven),
Leuven, Belgium
e-mail: margo.annemans@asro.kuleuven.be

A. Heylighen
e-mail: ann.heylighen@asro.kuleuven.be

C. H. Van Audenhove
LUCAS Center for Research and Consultancy in Care, University of Leuven (KU Leuven),
Leuven, Belgium

M. Annemans · H. Vermolen
Osar architects nv, Antwerp, Belgium

P. M. Langdon et al. (eds.), *Inclusive Designing*, DOI: 10.1007/978-3-319-05095-9_12,   131
© Springer International Publishing Switzerland 2014

1 Travelling Through the Hospital

> When you leave [the ward], and are [in the waiting area at the OR], you are mostly so
> nervous that you hardly notice anything, that you don't see things. There I was so nervous,
> I had never done this. [I thought:] Ooh, what can I expect? What am I going to feel?
>
> (Young woman about the route to the Operation Room (OR), 31 July 2012)

When undergoing a hospital admission, examination and operation, many people experience nervousness, agitation and uncertainty. This state of anxiety apparently increases once you are picked up to be transported to the OR. Buildings can influence people's experience positively or negatively (Day 2002), yet what does this mean in the context of hospital transport? What is the role of motion on your (spatial) experience of the hospital? Who and what plays a role in this experience? To start addressing these questions, we conducted a study at the day ward of a general city hospital. Since the length of hospital admissions is constantly reducing, more and more patients undergo treatment during a short, planned admission. Despite the diversity of this growing group of patients—some familiar, others unfamiliar with the hospital—they share a main concern with their health and the upcoming operation, combined with short contact times. Therefore we explored how we could make these patients reflect on their motional experience both during and after the transport, to and from the OR, while respecting the possible variation in their sensory, cognitive and physical capabilities and awareness. Reflections from this diverse group are expected to provide insight into how to design future hospital buildings that address hospital patients' needs before, during and after transport by focussing not only on functional and ergonomic aspects but also on people's personal experience related to their mindset under the given circumstances.

2 A Mobile Experience

While being transported through the hospital, to the OR, the bed is an important mediator between the patient and his or her environment (Annemans et al. 2011). Still the role of the bed in the general hospital experience is largely under-researched (van der Geest and Mommersteeg 2006). Research on beds and transport seems often restricted to functional matters such as organisation (e.g. College bouw zorginstellingen 2007) and ergonomics (e.g. Petzäll and Petzäll 2003; Kim et al. 2009; Mehta et al. 2011). Very little is known about patients' mobile experience in the bed.

When considering mobile experience it is important to distinguish between active and passive movement. Moving actively—whether as a pedestrian (Lorimer 2011), walker or wanderer (Lund 2005 in Paterson 2009; Bollnow et al. 2011), driver (Sheller 2004; Laurier 2011) or wheelchair user (Winance 2006)—is completely different from being passively moved as a passenger in a car or train (Bissell 2010), or being pushed in a wheelchair (Winance 2006) or bed. Since we can only feel transitions in movement (starts, stops, accelerations), we are unable to register a uniform motion in a straight line (Gibson 1968). If everything went perfectly smoothly the patient would not notice; what (s)he does notice is the change in direction or speed, resulting from bumps or starts and stops.

The negotiation between patient and world takes place at different levels, spatial, social and timewise. The bed mediates this relationship, thus heightening one's sensitivity to the building (Verbeek 2005). While you are lying in a bed, your various relationships with space are changed (Annemans et al. 2011). Your head-foot axis becomes horizontal, altering your entire system of perception. Your view of the environment is directed towards the upper part of the room and you register different haptic sensations through the bed's wheels, frame and mattress. Changing your posture from upright to lying down involves letting go of things, both literally and figuratively (Bollnow et al. 2011). When you are lying in a hospital bed, this letting go is even more emphasised by your loss of control on where you go and what is going to happen to you. As mentioned above, this interaction does not come into being because of the bed alone. As Myriam Winance (2010) illustrates for a wheelchair, the 'vehicle' itself, but also the one pushing it, play a crucial role in how the ride is experienced. Therefore, the research object is not just the 'person-in-the-bed', but the 'person-in-the-bed-pushed-by-someone-and-accompanied-by-someone-else'. Besides spatial and social aspects, an important element in mobile experience is the duration of the route. Speed is the distance travelled divided by the time of travel. However, as in the process of travelling multiple kinds of time and place are involved, socio-material practises have a major influence (Watts and Urry 2008). In a hospital, these could be fellow patients, spatial aspects like views or haptic sensations during transport or the sheets on the bed.

3 Towards a Mobile Research Approach

3.1 Researching Motion

Motion is intrinsically connected to space. So we looked for a way to gain a nuanced understanding of their meaning and mutual interaction. The key role of the environment and our focus on the subjective nature of human life—both the subjective experiences of the patients studied and the subjectivity of us, researchers—make a qualitative approach most appropriate for our study (Esterberg 2002). The specificity of the research adds some additional constraints. Involving experience and motion, the key role of the body and the difficulty of expressing spatial impressions verbally all present challenges to our research design.

Gaining insight into the embodied perception of patients being transported in a hospital bed requires a suitable research approach. Grasping all the sensual aspects of their movement requires an extension of traditional reflexive ethnography (Paterson 2009). Mobile research methods (doing research while being on the move) can stimulate encounters and communication (e.g. Pink 2008; Ross et al. 2009). Combining a dense description of your own experiences as a researcher, with the participants' stories and with photographs, audio and video recordings made by and alongside the participants, generates very rich sensory data, providing nuanced insights about both experience and physical aspects.

Due to ethical or other restrictions, not all contexts are suitable to actively engage participants in the making of photos or videos (Annemans et al. 2012). However, videography recorded by the researcher can also evoke in people clear reflections on a previously embodied experience (Merchant 2011). Therefore, we should not feel restricted to one on one recording, i.e. showing participants an actual recording of their own activity. Also recordings of similar activities made by others may trigger them to reflect on their own actions and experiences (Mollo and Falzon 2004). Based on these insights we modelled our research approach, adapted to the hospital context and balancing the stimulation of reflexivity in the participating patients with the recording of situation-specific, momentary data.

3.2 In the Field

Whereas a pilot study predicted promising results from the use of interactive visual methods (Annemans et al. 2012), the specific context in the day ward and the focus on the mobile aspect forced us to be flexible about the research set-up. Compared to other hospitalised patients those on the day ward spend little time in their room. Only very few turn out to have the motivation to consciously reflect on their environment and document this for use in a later conversation, possibly because of unfamiliarity with both location and circumstances. Therefore, although the objective of the research remained untouched, the research approach was adapted along the way.

Beforehand, a researcher (the first author) documented various aspects and spatial elements of the route. In an attempt to connect the researcher's and the researched embodied experience, a scenario was mapped out for registering visual, thermal and olfactory stimuli. As a start, the researcher herself took her place in a bed and was wheeled through the hallways to the OR and back. This trip was video recorded with a head camera and sensory perceptions were spoken out loud and audio recorded. For the entire route photographs were taken every five metres and every time the direction of the route changed or a new room was entered. At each point we planned on measuring the temperature. Every morning before starting the fieldwork, the researcher walked the route indicating clues to make up a smell map. Wear and tear in the hospital was photographed and indicated on a map.

The research was conducted over the course of 6 weeks, 1 day a week, at the day ward of a general hospital. During this period 12 patients were found willing to partake in the project. As agreed with the hospital's ethical board, a standard procedure started with a staff member asking a patient whether (s)he would like to participate in a study on spatial experience. If they agreed the researcher entered the room, and explained in more detail the content and goal of the research and what the participant could expect. If (s)he still agreed, an informed consent form was signed by both parties. Then the participants were given a photo camera, pen, pencils and paper, and asked, while they were waiting to be brought to the OR, to document their stay at the hospital. Although some did, many responded that they were not in a state of mind to do so. Also depending on the patient's psychological state, a short interview about the first impressions of the hospital was already conducted. Then the researcher left, until the patient was picked up by a logistic assistant to be wheeled to the OR. These transports were accompanied by the researcher, by analogy with guided walks (Pink 2008; Ross et al. 2009) and audio recorded. No specific questions were asked but it was made clear by the researcher that patients were free to comment on their trip (talk out loud) or stay silent. The same was done for the route back to their room. When they were wide awake again, the actual interview took place. The initial intention was to base this on the documentation provided by the patients. However, for both personal and practical reasons only a limited amount of this information was available. Additionally, as they were preoccupied with the upcoming operation when going, and sometimes quite sleepy when returning, few had a clear memory of the transport. It soon became clear that watching the video of the researcher's travel helped them to reflect on theirs as well. As such, this video material became an important element in the interviews.

The audio recordings of both the accompanied transports and the interviews were transcribed verbatim. The transcripts, complemented with visual material when available, were analysed through coding them using qualitative data analysis software (ATLAS.TI). To establish a list of representative codes, we started from previously established categories grounded in literature (Creswell 2003) and expanded the spectrum by in vivo codes expressing topics that emerged through the process.

4 Unravelling Mobile Experience

4.1 Space

Some sensory perceptions can be directly linked with spatial elements like lights, joints, ceilings, walls or furniture. Several patients experienced the sequence of the lighting combined with the dull ceiling as very unpleasant. As one stated: *From the moment I left here I saw those lamps pass by one after the other. That was the worse. I hate that. It is like a movie that you see when you go to the OR, where you cannot get out.* Others express it in a more positive way, suggesting changes to the ceiling, or the upper part of the wall. *Yes, they could probably do something about that, I don't know, maybe the colour or something like decoration, so that it would be a little more pleasant to ride there.* Besides the visual also the haptic experience was discussed frequently. Mostly the bed itself is considered comfortable and got lots of compliments on its ability to be adjusted to one's personal needs. However, when being transported in a bed, each unevenness during the ride is felt. Bumping into things can cause these interruptions but also joints in the floor, a door profile, and the entrance of the elevator interrupt the smoothness of the ride. It all adds to a participant sighing: *It seemed like a cobble stone street, all those bumps.*

Other perceptual experiences, like temperature and sound, are a more indirect consequence of spatial or material aspects. The temperature in the building is influenced by basic architectural decisions, e.g. the ward's orientation, the floor on which a room is situated or the presence of (open) doors and windows. At the time of the fieldwork the weather was fairly warm, so the difference in temperature between the ward and hallway downstairs close to the OR was significant. However, this was not necessarily experienced as negative. When talking about the difference between active and passive movement, a patient mentioned: *[what I never noticed (when actively moving) is] the difference in temperature. When you go downstairs it is colder again. [On the way to the OR] we passed by a door and that [the presence of fresh, cool air] stays with you. It gives some variety. [That is important] because when you are lying, you don't see these things.* This illustrates the complementarity of different senses: while being transported in a bed your visual perception may decrease, whereas your sensitivity to other sensory experiences, like feeling the fresh air, could increase. A material element can also have a direct and indirect impact. The bumps mentioned above result in a disturbing feeling, yet also the sound generated is found 'typical' for hospital transport. The audio recordings of the transport confirm the presence of this sound. Frequently heard in the interviews were quotes like *It is like that when you are brought there, over all those bumps tum, tum, tum, tum, yes that is* and *those ridges here, that is not pleasant when you go that is not a problem, but when you return, with that head, constantly the click, click, click, click. That is not pleasant.*

The way participants phrase their answer already highlights the distinct rhythm as a very specific attribute of the transport. Also distance and direction fit in this segment. The length of a trajectory is perceived differently because of the lack of control and the difference in speed. The direction in which a patient is transported, head first or feet first, depends on the type of bed and the preference of the person doing the transport. Although related to space, both are tightly connected with time and/or people, discussed in the following paragraphs.

4.2 Time

How far one place is from another is clearly a spatial matter. How far it seems to be, however, has a lot to do with how smoothly and, even more important, how fast the ride goes. Whether you like being transported apparently is a very personal issue. Still, the speed of the ride seems to be a significant factor in appreciation. Some patients were really bothered by the rhythm generated by the moving bed related to the spatial aspects discussed above. Others said things like: *Now I have to rest. I found that a nice distraction, being wheeled in the bed, they should have moved around some more with me.* When the interviewer asked whether the route should have been longer, the answer was: *Yes, but sometimes they rode pretty fast. I liked that. I am lying in this bed, so nothing can happen to me.* This man considered the ride a nice distraction. The speed of the ride contrasts with the rest of the time spent in the day ward where you are actually just waiting for the next thing to happen. At the same time, the bed's presence provides a feeling of safety. Although as a patient you are well cared for, you spend a lot of time alone, whereas during the transport there is always someone accompanying you. Apparently this gives the feeling that nothing can go wrong.

The value of the movement time becomes even clearer when we take a closer look at its counterpart, the waiting time. While waiting you are not actually moving, yet it can be considered an important part of the route to and from the OR that is studied here. How long patients had to wait in a certain place or at a certain point along the route was frequently mentioned. A couple both of whom were admitted for an examination held the following conversation: *I had to wait a long time there before it was my turn. Did you? Yes I did, yes.* Then she turned to the researcher saying: *That is because he was the first and I was the last one, and then you don't know anything of him and ooh, I am then a little … [nervous].* The lack of control over the situation and not knowing what to expect next seems to add to the feeling of time passing very slowly.

4.3 People

During the entire route from the ward to the OR, you are hardly ever on your own. Various people interfere with you at different stages of the trajectory. When coming to the hospital, a parent, child, partner, or someone else may accompany you. In the hospital room, you may have a roommate. Even if you do not, nurses come in every now and then to check on you or prepare you for your examination. Once a logistic assistant picks you up, you leave the people familiar to you but are still with someone. Along the ride, others pass by in the hallway and once your bed is parked at the waiting room in front of the OR, many others stand next to you. All these people influence your experience, whether physically or through verbal interaction.

Obviously, the logistic assistant pushing the bed has a direct influence on the motional experience. (S)he decides whether to push the bed from the head or the feet, depending on the type of bed, own preference, possible obstacles on the route, and general instructions. As such, (s)he also determines the direction you will travel, head first facing backwards, or feet first facing in the same direction as you ride. One participant said: *[…] because they ride with your head first, if they would ride with your feet first, but now you think: Ooh, what will be coming? What will be coming?* The video material was recorded moving feet first, while most patients were transported in the opposite way. Many comments about what they saw related to the difference between both perspectives, mostly stating that they saw a lot less. Not knowing what will come or what is happening around you, you have to let go of the control you have when moving actively. A woman characterised the difference as follows: *It is nicer, when you walk, you have more control. It is always like that. When you lay in a bed then you are entirely in the hands of the [nurses] [...] and then you have to ... yes, follow. That is different from when you walk, then you determine your own rhythm and you see different things too.*

There seems to be a constant interaction between how you experience others' presence around you while moving and how you experience space (or spatial aspects). In the (multiple) patient rooms, in the waiting and recovery rooms, curtains between beds provide at least some visual privacy. However, when they are closed, the patient at the side of the hallway is left in the dark, especially in the rooms. In the waiting and recovery room, open curtains leave other patients and possible medical interventions in full view. As a participant said: *[...] it can be that you are afraid of the operation or the unknown. Then it can be a little frightening when you are standing there in the waiting area, because you are there with other patients, and you see all those people, although there is a little curtain, but you see them pass by and if someone comes there for the first time. You see the doctors and the nurses with their mask and their head and the OR clothes. Yes then it is possible that you say 'Ooh, no, what is happening here'* Closing the curtains in the waiting room or recovery implies no or less visual surveillance of the nurses over the patient, but also less contact of the patient with the staff. Apart from a possible visual separation, neither patients nor staff have much privacy, especially in the waiting area. Sometimes overheard conversations can cause irritation. A former nurse who now participated in our study as a patient testified: *When I entered the waiting area of the OR, I thought it was horrible. It was like a nice clique. They were making paella and someone had holiday pictures with her. That is allowed, but as a patient I find that disturbing.* In situations like this, rethinking the spatial organisation could significantly contribute to decreasing unnecessary but very human irritations.

5 Discussion

The study presented here is part of a broader research project. Also the fieldwork conducted at the day ward covered a much wider spectrum of experiences of the hospital stay. For the analysis reported here we focused on the area of transport. Spatial aspects related to the hospital stay, in the ward or a treatment room are not addressed, if not related to the route followed. However, this does not mean that the participating patients limited their responses to this route. Many of them reduced the importance of the spatial experience by pointing to the limited duration of their stay. Yet, all of them agreed that a supportive environment would be much more valued when they had to stay longer than a day. This is being investigated in another part of our research.

Various patients also questioned the necessity of being wheeled to the OR while being perfectly capable of walking there. Although not all agreed that it would be better, comparing the experiences of the patients participating in this study with those of patients in a walk-in-day-ward could be interesting, and is taken up in the further development of our project. The participants in this study confirmed a significant difference between actively moving, while walking and being transported in a bed. As the literature on mobile experience focuses largely on active movement, the insights collected here add to the knowledge of its passive counterpart. Since most of these studies look at mobility on an urban scale, translating their results to a building scale presents a major challenge. Still, we could also draw significant parallels. Socio-material practises (Watts and Urry 2008) are indeed a major influence, relating the three topics of space, time and people. Also the extended research object (Winance 2010) is confirmed and even further extended into the 'person-in-the-bed-pushed-by-someone-and-accompanied-by-someone-else'. However, whereas in the case of a 'wheelchair-user-unity', the unity is (mostly) fixed and familiar, in the case of the bed, none of the actors are familiar with each other, which could add to the feelings of uncertainty and loss of control. As the analysis shows, clearly distinguishing between the categories is hard. All are intertwined and constantly interacting. Unravelling them is challenging.

Also the adopted research approach deserves a closer look. Reflecting on your spatial experience is not easy. Additionally, the patients' experience was influenced by many others interacting with the transport. Accompanying relatives, commenting during the interview, or pointing at different things, or logistic assistants warning the patient to cover up well because of the difference in temperature may have influenced the responses. By making use of the video material the researcher's perspective sneaked in as well. Although we aimed to make the recording as neutral as possible, the researcher looked at certain things, and the one pushing the bed, did so in a certain way, feet first, sometimes trying to avoid obstacles that were absent when the participants passed the same place. However, most participants were able to set aside these differences and relate the recording to their own experience, pointing at parallels or differences and as such providing additional data and confirm earlier research (Mollo and Falzon 2004). Also the extra information, unasked for, given by others, can be understood in this perspective. As illustrated, a real-life hospital experience is shaped by many. Therefore, all in all, we believe that, given the specific circumstances, the approach adopted found a good balance between giving people the freedom to respond at the moment they feel most comfortable and address the topics relevant to them and relating their responses to the actual environment.

6 Concluding Remarks

Studying the role of motion in patients' spatial experience of a hospital is not easy. Both methodologically and content-wise, we encountered some challenges in bringing to the surface and unravelling those experiences linked to spatial aspects. Although at first sight one could expect spatial aspects to disappear to the background when being transported in a bed, e.g. due to the reduced view of the environment, this study shows the opposite. Indeed, the sensory realm is broadened by new perceptions. A great challenge for architects lies in translating this kind of findings into a positive stimulus for patients.

Spatial aspects cannot be studied in isolation from the activities going on in the spaces considered. Depending on what is happening and who is involved, each experience is shaped differently. Also during transport, we encountered an important interrelation between space, time and people. As such, architects designing space also strongly influence time perception and people's interactions. Since particular spatial elements can have only a momentary impact on these fields, we should aim to approach the design of health care buildings in a more general way, where architecture and organisation go hand in hand. Although studying organisational matters transcends the scope of this research project, we are convinced that, due to the connecting character of motion, interesting insights concerning spatial relations could be derived, especially when broadening our horizon, not just focussing on one route, but combining research on different routes and modes of transport.

Acknowledgments Margo Annemans' research is funded by a PhD grant of the Baekeland programme of the Institute for the Promotion of Innovation through Science and Technology in Flanders (IWT-Vlaanderen), which gives researchers the opportunity to complete a PhD in close collaboration with industry, in this case, osar architects nv. Ann Heylighen received support from the European Research Council under the European Community's Seventh Framework Programme (FP7/2007-2013)/ERC grant agreement n° 201673. The authors thank the participating patients for sharing their time and insights and the hospital board for their support.

References

Annemans M et al (2011) Lying architecture. In: Proceedings of the WELL-BEING 2011, Birmingham, UK
Annemans M, Van Audenhove C, Vermolen H, Heylighen A (2012) Hospital reality from a lying perspective. In: Langdon PM, Clarkson PJ, Robinson P, Lazar J, Heylighen A (eds) Designing inclusive systems. Springer, London
Bissell D (2010) Vibrating materialities. Area 42(4):479–486
Bollnow OF, Shuttleworth C, Kohlmaier J (2011) Human space. Hyphen Press, London
College bouw zorginstellingen (2007) Patiëntenstromen en zorglogistiek in het ziekenhuisgebouw. College bouw zorginstellingen, Utrecht
Creswell J (2003) Research design, 2nd edn. Sage, Thousand Oaks
Day C (2002) Spirit and place. Architectural Press, Oxford

Esterberg KG (2002) Qualitative methods in social research. McGraw-Hill, Boston

Gibson J (1968) The senses considered as perceptual system. Allen and Unwin, London

Kim S, Barker LM, Jia B, Agnew MJ, Nussbaum MA (2009) Effects of two hospital bed design features on physical demands and usability during brake engagement and patient transportation. Int J Nurs Stud 46(3):317–325

Laurier E (2011) Driving: pre-cognition and driving. In: Cresswell T, Merriman P (eds) Geographies of mobilities. Ashgate, Farnham

Lorimer H (2011) Walking. In: Cresswell T, Merriman P (eds) Geographies of mobilities. Ashgate, Farnham

Lund K (2005) Seeing in motion and the touching eye. Etnofoor 181:27–42

Mehta RK, Horton L, Agnew MJ, Nussbaum MA (2011) Ergonomic evaluation of hospital bed design features during patient handling tasks. Int J Ind Ergon 41(6):647–652

Merchant S (2011) The body and the senses. Body Soc 17(1):53–72

Mollo V, Falzon P (2004) Auto- and allo-confrontation as tools for reflective activities. Appl Ergon 35(6):531–540

Paterson M (2009) Haptic geographies. Prog Hum Geogr 33(6):766–788

Petzäll K, Petzäll J (2003) Transportation with hospital beds. Appl Ergon 34:383–392

Pink S (2008) An urban tour. Ethnography 9(2):175–196

Ross NJ, Renold E, Holland S, Hillman A (2009) Moving stories. Qual Res 9(5):605–623

Sheller M (2004) Automotive emotions. Theor Cult Soc 21(4/5):221–242

Van der Geest S, Mommersteeg G (2006) Beds and culture. Medische Antropologie 18(1):7–17

Verbeek PP (2005) What things do. Pennsylvania State University Press, University Park

Watts L, Urry J (2008) Moving methods, travelling time. Soc Space 26:860–874

Winance M (2010) Care and disability. Practices of experimenting, tinkering with and arranging people and technical aids. In: Mol A, Moser I, Pols J (eds) Care in practice. On tinkering in clinics, homes and farms. Transcript Verlag, Bielefeld, Germany

Winance M (2006) Trying out the wheelchair. STHV 31(1):55–72

Student Engagement in Assessment of Universal Access of University Buildings

Z. Ceresnova

1 Introduction

Universal design (UD) is often considered by architects to be something ancillary to architectural design and focused only on a small group of users. But it is an integral part of the design process that aims to create a better environment for all people.

> Contrary to the assumption that attention to the needs of diverse people limits good design, the results of imaginative designers around the world reveal a wide range of applications that delight the senses and lift the human spirit when "universal design" is integral
>
> (Ostroff 2011).

Students of architecture should be educated to be more empathic in taking the diversity of users of the built environment into consideration.

The article presents the research and the educational project conducted by Associate Professor Maria Samova and the academic staff of the Centre of Design for All (CEDA) at the Faculty of Architecture, Slovak University of Technology (FA STU) in Bratislava. The title of the project was: Universal Design of the Built Environment Focused on the Inclusion of People with Disabilities into the Education and the Work Process. We engaged our students to take part in the various phases of the project. One of their tasks was to observe and evaluate selected university buildings from the position of users with disabilities.

Z. Ceresnova (✉)
Centre of Design for All, Faculty of Architecture, Slovak University of Technology
in Bratislava, Bratislava, Slovakia
e-mail: ceresnova@fa.stuba.sk

P. M. Langdon et al. (eds.), *Inclusive Designing*, DOI: 10.1007/978-3-319-05095-9_13,   143
© Springer International Publishing Switzerland 2014

2 Towards Universal Design in the Educational Environment

The aim of UD is to create a non-discriminatory environment. The inclusion of people with disabilities into everyday life should not be only wishful thinking—these people have to have an opportunity to gain an education and be part of the working process just like other groups (Ceresnova 2013). The right to education can be successfully implemented only if educational facilities are designed to give everyone an equal opportunity to reach, enter and use all the facilities (Heylighen et al. 2006).

The ambition of UD is to combine contextual information, for example, the characteristics of users and the built environment so that the design of the education and the work environment is convenient for diverse users (Samova et al. 2006a, b). The understanding and correct application of accessibility principles is not possible without getting familiar with users' requirements. The pedagogical aspect of realised physical accessibility in the educational facilities is also important, as Heylighen et al. (2006) pointed out *young people are educated as much by example as through teaching. Environments that segregate teach acceptance of segregation, and inclusive environments teach inclusion.*

UD is the methodology for creating an environment that achieves the inclusion of people with disabilities into society. Froyen et al. (2009a, b) defines UD

> to be an academic and a professional research and design response to a democratic requirement and to a social concept for the achievement of integral and inclusive accessibility and utility.

An inclusive educational environment is related not only to physical accessibility, but also to the accessibility of teaching and learning methods for diverse users. Methodology of universal design for learning (UDL) developed by Rose and Meyer is focused on the implementation of three main principles in the educational process: (1) multiple means of representation, (2) multiple means of expression, (3) multiple means of engagement (Burgstahler and Cory 2008). Teachers should present materials to students in multiple ways and students should have the possibility of demonstrating their knowledge in multiple formats (e.g. multimedia presentations, written papers, photo documentations, drawings, and sketches). The engagement and the motivation of students are very important for effective teaching (Hehir and Katzman 2012). Therefore, the teaching of UD should concern the active engagement of students. Role-play activities and on-site observation can help to better understand users' needs and the principles of UD (Samova et al. 2006a).

2.1 Student Engagement and Their Tasks

Our research was closely connected with teaching of the students of architecture. In the project, we involved students from different years of their study—from the second to the fourth year. They participated in various tasks within the compulsory subjects design studio (DS) and UD.

There were 195 students involved both individually and in pairs in the first phase of our project that took place in the winter semester 2011 within the subject UD. Students' task was to map the current state of accessibility of primary school buildings across Slovakia. The partnership with the schools was created through their former pupils, who are the students of our faculty now. Our students come from various towns and villages of Slovakia; it enabled us to collect data from different parts of our country. We mapped approximately 150 school buildings. The aim was to analyse the obstacles in the school environment and then to propose solutions to eliminate the barriers. Some students used the interactive form of the on-site survey with the active participation of the pupils. These students conducted the survey together with a selected volunteer from the pupils who was seated in a wheelchair. The point was that the difference between a pupil and an adult on the wheelchair is evident, especially in their reach distance, and in their power to move the wheelchair independently. So, it was more realistic to map the environment from the position of a child in a wheelchair.

The second phase of the project was connected with the subject DS in the spring semester 2012. Approximately, 100 students received the task of designing the prototype of an inclusive school building for primary education in the new residential area in the suburban part of Bratislava. The aim was to create an accessible and friendly environment for 260 pupils including the children with disabilities.

The most active engagement of the students was evident in the third stage of our project within the subject UD in the winter semester 2012. The total number of students involved in this phase was 151. To make the education more effective, we applied the principles of UDL. Therefore, each student could choose the task and the form of the elaboration and thus also choose the type of final evaluation. The main task was focused on the accessibility of university buildings and campuses in Slovakia and abroad. There were two subtasks to be chosen: (1) the analysis of selected buildings of foreign higher institutions or (2) the accessibility survey of our university buildings in Bratislava. Depending on the selected subtask there were two forms of elaboration: (1) an analysis in written form with photo documentation worked out individually by the student, (2) a survey in the form of a video document created by a group of three students. The survey was carried out from the position of the users with different disabilities—physical and sensorial. The evaluation process of students' knowledge comprised two forms: (1) the students who elaborated the first subtask had to pass a written test on UD principles and (2) the students who did the second subtask had to prepare a presentation on how the building from their video document met the criteria of UD principles.

The fourth stage of our project took place within the subject DS during the spring semester 2013. The task for approximately 70 students was to design a new Inclusive Kindergarten or a University Incubator at the university campus in Bratislava. The kindergarten should be designed to accommodate children with disabilities together with other children. Therefore, it was necessary to create appropriate space for indoor and outdoor activities, children's games and also accessible hygienic facilities in each department. The university incubator should allow young people with disabilities to work in flexible offices using all services without any restrictions.

The involvement of the students in various phases of our project was focused not only on the assessment of existing educational facilities, but also on the design of new inclusive settings. The students were encouraged to be creative in designing their proposals for new inclusive school facilities for various educational levels from kindergartens to higher education institutions. The different stages of education were considered in our research, because inclusive education should be implemented at every level, starting from early childhood. We also concentrated on the selection of buildings in different conditions and states. We were mapping: (1) buildings with overall accessibility adaptations, (2) buildings with partial or absent accessibility interventions and (3) newly constructed buildings which should be accessible according to legislation. We compared and evaluated the differences in these three sorts of university buildings.

The activities of our project consisted of different kinds of research tasks with the aim of designing and also evaluating physical environments:

> *Designers must not only research and incorporate suitable facilities for users with permanent or temporary physical and/or mental functional limitations, but they must also focus on the much broader totality of handicap situations. These handicap situations are not necessarily medically-related but are often a consequence of poorly adjusted designs*
> (Froyen et al. 2009a, b).

2.2 Examples of Assessment of Universal Access

Three selected examples of university facilities are presented documenting different qualities of accessibility: (1) a building with overall accessibility adaptations, (2) a building with partial interventions, and (3) new building which is supposed to be constructed without barriers according to the legislation. The students carried out the video and photo-analyses of selected higher education institutions. They mapped and documented the problems and conflicts between the users and the built environment (Froyen et al. 2009a, b) and then they made some recommendations for solving existing negative situations. The students assessed the buildings from the position of users with different disabilities—physical and sensorial (Preiser 2008). This teaching tool of UD is very often used in Belgium:

> *by simulating a diversity of physical and mental disorders, the students develop the ability to appraise whether a designed environment tends to create or eliminate handicaps*
> (Dujardin 2011).

The students were allowed to choose the type of disability for the simulation exercises. They preferred to move in a wheelchair. A small number of students chose to simulate blindness with special darkened glasses. Introductory training on spatial orientation with a white cane was necessary for the simulation of a blind person's movement. However, not every student completed this training, so their movement in the environment was not completely identical with the real movement of blind persons.

This active way of acquiring knowledge was very inspiring for both students and teachers who were involved in this process.

2.2.1 The Faculty of Civil Engineering STU in Bratislava (Building with Overall Accessibility Adaptations)

The first group of students assessed the building of the Faculty of Civil Engineering, built in the years 1964–1974. Overall accessibility adaptations to the building were realised in 2010. At the beginning many people claimed that too much assets had been spent only for one person—at that time only one student with mobility impairment studied there. But the concept of the new interventions was to facilitate easy movement within the building for everyone. In the past, there were only two main staircases in the entrance hall leading separately into the two wings of the building without any accessible connection between them. People had to go downstairs and then upstairs, but now there is the new footbridge helping to connect these two parts of the building.

The three students of the group divided their roles: the first student acted as a person in a wheelchair, the second one as a person with visual impairment with a guide dog and the third one filmed with the camera. The on-site observation started at the main entrance, accessible by a newly constructed ramp. In accordance with the legislation, there were tactile guiding lines at the platform leading to the main entrance. The guiding lines enable people with visual impairment to identify the entrance door and then they continue to the entrance hall with reception desk. Students pointed out that it was easy to identify the entrance and the reception desk, but the rest of the building was complicated for orientation, because there was no information in multisensory form or clear guiding lines.

The students continued their survey in the canteen. They found some problems there, such as the narrow door, the inaccessible toilet and not clearly identifiable room layout. Then students visited the Sports Centre with the swimming pool, the fitness area and the gym. Access to the Sports Centre is only via staircase; it means that the student in the wheelchair was unable to go there. It is a basic problem that the sport activities are often omitted from the accessibility concept, because many architects think that these activities are not for people with disabilities, and legislation is also insufficient in this field.

Our student actors attended some lectures in the main auditorium. They could enter the room, but there was no space for users in a wheelchair to stay behind the table desk, only to be separated in front of the tiered seating.

The frequent mistake of architects is that they solve the access to the room, but they do not take care to enable inclusive usage of the space. People can enter, but they cannot use the space appropriately (Fig. 1). It was visible in many situations tested by our students, for example, the Internet kiosk placed out of reach of the person on the wheelchair. And also the reception desk was too high for them.

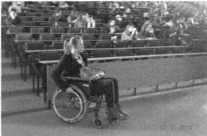

Fig. 1 The Library and the Auditorium at the Faculty of Civil Engineering STU (accessibility survey conducted by students)

The students identified various types of hazardous situation, for example, one ramp with the low headroom height. Maybe it was anticipated that the ramp would be used only by people in wheelchairs who are not so high when sitting. But the problem would occur for the person accompanying the wheelchair user. Nevertheless, it is also discriminatory that the ramp cannot be safely used by everyone.

2.2.2 The Faculty of Mechanical Engineering STU in Bratislava (Building Without Complete Accessibility Adaptation, Only Partial Adaptation)

Another group of students selected the building of the Faculty of Mechanical Engineering built in the years 1948–1963, which has only partial accessibility adaptations. This group of three students chose a different approach to surveying the accessibility of the object. The scenario was based on the story of two students in wheelchairs, who wanted to meet at college—in the cafeteria; each of them came from a different way. One of them used a car, which was parked in the courtyard of the building and came to the building using the side entrance near the parking lot. The second one came by bus and got off at the bus stop near the main entrance of the building. Both people were trying to get to the cafeteria, but finally, the meeting was not easily feasible even though they used the partial accessibility adaptation of the building. There was no overall accessibility concept there and hence some stairs created the barriers between the different parts of the building.

2.2.3 The Faculty of Informatics and Information Technologies STU in Bratislava (the New Building, Constructed in 2012)

The new building of the Faculty of Informatics and Information Technologies was opened in 2012. Students who made the accessibility survey of the building expected perfect accessibility solutions. They started their route from the point of transport accessibility—to find out the possibilities for the person in the wheelchair to get to the bus and then from the bus stop to the building. The students pointed out that the route from the bus stop to the main entrance was very difficult due to the steep terrain and long walking distance. The main entrance was not easily identifiable and, moreover, it was situated in the furthest position from the bus stop. There were also some side entrances along the way, but they were inaccessible due to many stairs.

Inside the building there was a problem with opening some doors with the self-closing system. A very dangerous situation occurred in the toilet cabin; when the self-closing doors closed quickly, suddenly it was dark inside, while the light switches were placed out of reach. So the person in the wheelchair found himself in the position of a blind person for a while.

2.2.4 Assessment of the Surveys

The common feature of all buildings assessed was that the principles of UD had not been applied in those elements and spaces which are not explicitly defined in building codes (e.g. the height of counters, the seating solutions for wheelchair users in the auditorium, access to the library shelves and info boxes). Also, the accessibility of some activities (especially sports and leisure activities) was not resolved because mainstream thinking excludes persons with disabilities from participating in various activities.

Comparing the participating groups to each other, we found different student perspectives on physical accessibility and different priorities in accessibility assessment. Some groups addressed accessibility using public transport but the object itself was not mapped in detail (half of the presentations concentrated on the survey of transport accessibility more than the facility itself). Other groups focused only on the object itself from its main entrance through each space of the building (the auditorium, the gym, the library, the cafeteria, the services, etc.). Also the viewing angle and the filming were different—most of the groups elected one person as an independent cameraman, while the two others were actors in the movie. Only one group from a total of 10, filmed directly from the wheelchair with a camera mounted on it. That really gave us an interesting story.

2.3 Pedagogical Aspects of the Project

The project was implemented in two compulsory subjects: (1) UD and (2) DS. The subject UD was taught only by experts from the Centre of Design for All at our faculty and partners from the organisations of persons with disabilities. The teaching of DS was done by all the teachers of our department, for many of them the topic of UD was not well known. After completing the introductory lectures and consultations with experts, the teachers did become more familiar with UD.

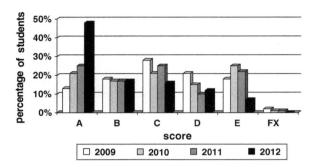

Fig. 2 Comparison of the students' evaluation results in UD in years 2009–2012 (scores from grade A Excellent to grade FX Insufficient)

That way we achieved good dissemination and raised awareness of accessibility issues among all the teachers. They were, thus prepared to advise students also on how to implement the principles of UD into their design proposals.

The final evaluation of students' knowledge gained in the subject UD showed the effectiveness of the UDL method (by using multiple means of representation, expression and the active engagement of students) in comparison with traditional teaching methods. According to the principles of UDL, we offered the students the opportunity to demonstrate their knowledge by multiple ways in the final evaluation in winter semester 2012. Students had two options (in relation to their elaboration of subtasks): (1) a written form of the test on UD principles and (2) an oral presentation of the assessment of the university building in accordance with UD principles, pursuant of its on-site survey. Both forms of examination were evaluated by three teachers using the scoring system from grade A (excellent) to grade FX (insufficient).

Based on the results of the final evaluation, it was found that active engagement of the students helped to improve their understanding of the principles of UD. Comparing the results received from the academic information system of STU in Bratislava for the period from 2009 to 2012; it is evident that better results (score A) were achieved by fostering the principles of UDL in the education in the academic year 2012 (Fig. 2).

3 Conclusion

On-site observations and role-play activities can help students to think about architecture more sophisticatedly—by experiencing from the position of users with disabilities how they perceive space and how they are able to use it. For ordinary people, some of the problems are not so visible, or identifiable. It is very useful for each designer and architect to get under the skin of various kinds of users to understand better the requirements and needs they have and to aim for the creation of the inclusive environment.

Our research project highlighted measurable streamlining of teaching UD through the implementation of UDL. Further, important contribution from the project was the development and formulation of principles and recommendations for design of the inclusive educational environment based on surveys of existing buildings. The output of the project will help architects, civil engineers and designers to create better environments allowing access and appropriate conditions for the education of people with disabilities, ensuring their inclusion into society. The results of this research project could be a stimulus for the revision of the current building legislation in the Slovak Republic in accordance with the requirements of the European Union to realise the equal right of every citizen to education.

Acknowledgments The project reported in this paper received funding from The Scientific Grant Agency of the Ministry of Education, Science, Research and Sport of the Slovak Republic and the Slovak Academy of Sciences, grant agreement No. 1/0996/11. The author thanks all the people who participated in this project, in particular the students of architecture and the academic staff of the Centre of Design for All (CEDA) at the Faculty of Architecture, Slovak University of Technology in Bratislava.

References

Burgstahler S, Cory CR (eds) (2008) Universal design in higher education: From principles to practice. Harvard Education Press, Cambridge

Ceresnova Z (2013) Universal design of inclusive learning environment. In: Universal learning design. Proceedings of the conference. Masaryk University, Teiresiás, Brno

Dujardin M (2011) Learning from practice: Post-occupancy evaluation (POE) as a UD teaching tool at Saint-Lucas Architecture, Belgium. In: Proceedings of the Include 2011, Helen Hamlyn Centre, London

Froyen H, Verdonck E, De Meester D, Heylighen A (2009a) Documenting handicap situations and eliminations through universal design patterns. Australas Med J (AMJ) 1(12):199–203

Froyen H, Verdonck E, De Meester D, Heylighen A (2009b) Mapping and documenting conflicts between users and built environments. In: Proceedings of the Include 2009, Helen Hamlyn Centre, London

Hehir T, Katzman L (2012) Effective inclusive schools. Designing successful schoolwide programs. Jossey-Bass, San Francisco

Heylighen A, Michiels S, Van Huffel S (2006) Towards universal university buildings. In: Universal design of buildings: Tools and policy. Proceedings of the POLIS International Conference, Bruges, Belgium

Ostroff E (2011) Universal design: An evolving paradigm. In: Preiser WFE, Smith KH (eds) Universal design handbook, 2nd ed. McGraw-Hill

Preiser WFE (2008) Universal design: From policy to assessment research and practice. Archnet-IJAR Int J Architectural Res 2(2):78–93

Samova M, Ceresnova Z, Rollova L (2006a) Design for all implemented in curricula using different tools. In: Proceedings of the BAS international conference, Budapest, Hungary

Samova M, Ceresnova Z, Rollova L, Korcek P, Majcher S (2006b) Architecture and urban planning. Design for all from school to practice. ARVHA, Paris

Reported Design Processes for Accessibility in Rail Transport

R. Herriott and S. Cook

1 Introduction

Accessibility is a fundamental requirement in public transport (PT) yet, there exists little research on design for accessibility or inclusive design (ID) in this area. This paper sets out to discover what methods are used in the rail sector to achieve accessibility goals and to examine how far these methods deviate from user-centred and ID norms. Semi-structured interviews were conducted with nine rolling stock producers, operators and design consultancies. The purpose was to determine if ID design methods are used explicitly and the extent to which the processes used conformed to ID (if at all). The research found that the role of users in the design process of manufacturers was limited and that compliance with industry standards was the dominant means to achieving accessibility goals. Design consultancies were willing to apply more user-centred design if the client requested it. Where operators were in charge of the design process, accessibility was addressed more comprehensively, with mixed results. The work suggests that the more the design process is divided among actors and geographic space, the harder it is to integrate users.

2 Background

Accessibility is essential if PT is to achieve the goal of serving citizens equally. There are provisions in the UK's Equality Act relating to accessible public service vehicles such as buses, coaches and rail vehicles (HM Government UK 2010). The European

R. Herriott (✉)
Design Platform, Aarhus School of Architecture, Aarhus, Denmark
e-mail: richard.herriott@aarch.dk

S. Cook
User-Centred Design Group, Loughborough Design School, Loughborough University, Loughborough, UK

P. M. Langdon et al. (eds.), *Inclusive Designing*, DOI: 10.1007/978-3-319-05095-9_14,   153
© Springer International Publishing Switzerland 2014

Union has drafted the Technical Specification of Interoperability—Persons of Reduced Mobility (EU Commission 2008), which specifies standards for visual equipment, boarding aids and tactile signage among other features. In this context, ID would appear to be a suitable means to maximise accessibility in public transport: it is user-driven (British Standards Institute 2005) and focuses on the needs of the broader mainstream of passengers. But little research exists related to ID methods and PT, particularly rail, despite the accepted necessity of accessibility in that context (Tennoy and Lieren 2008). This point is returned to in the literature review.

The value of accessible PT design can be explained as follows: 'Mobility declines with increasing age, reflecting the onset of physical or mental infirmity, and reduced affordability of travel for those on retirement incomes' (Metz 2003). Exclusion from PT diminishes quality of life (Social Exclusion Unit 2003). Additionally, disabled citizens of working age can be excluded from full participation in economic life if their choice of employment is limited by barriers to mobility (Kitchin 1998; Wilson 2003). The problems encountered are not only physical but also psychological (Hirsch and Thompson 2011). But not all PT is equally unsatisfactory. 'In terms of convenience and ease of use, taxis and minicabs are rated the most highly, with rail services the worst' (Miller et al. 2006). Thus, rail warrants a special investigation. Barriers in PT exist but not just because 'that's the way it is'. It is because the designers and manufacturers have made it so. This paper sets out to find out what methods are used in the rail sector to achieve accessibility and to examine how extensively these methods deviate from user-centred and ID norms.

3 Literature Review

There appears to be a lack of research in the area of design for public transport as it concerns rail passengers' needs. Such material as exists is scattered around related areas and not easily accessed. The material found can be divided into three groups. One relates to passenger ergonomics but does not focus on accessibility (e.g. Wilson et al. 2004; Davis and Mills 2005; Hirsch and Thompson 2011; Lauren et al. 2005). A second relates to accessibility in transport but does not cover rail vehicles (e.g. Azmin-Fouladi 2007; Baird and Dong 2009; Kaneko et al. 2009; Nickpour et al. 2012). A third, smaller group is concerned with ID in travel but rather than describing cases or practice, makes recommendations for how to improve the accessibility of the designs (e.g. Langdon et al. 2009). Evans (2005) described a survey of 'drivers and blockers' of accessible PT projects of which rail was one of many aspects. Marshall et al. (2009) approach the theme most closely in an innovative project. Their work involves a CAD-based system for providing usable ergonomic data for designers and a means for using the data in validation. However, the project concerns data rather than actual users and, when applied, serves as a proxy for user validation. Mackie (2012) and Woodcock (2012) provide an over-view of user-centred design methods and objectives in public transport.

Another way to summarise the literature is: ethnographic work and surveys have been done with rail users but not on matters of accessibility and not with express regard to elderly and disabled rail users; methods for assessment have been proposed but did not focus on rail; recommendations exist for how to get designers to think more about accessibility but this is not the same as examples of practice. This paper thus addresses a gap in the knowledge concerning design for accessibility in rail transport.

4 Methods

Designers at nine rail manufacturing firms, operators and design consultancies were interviewed. To arrange these interviews, it was necessary to contact 18 organisations working in the area of design for rail transport. A total of 114 e-mails or telephone calls were made from October 2010 to February 2012. Participants were asked to describe what steps they took, if any, to ensure their products were accessible for persons of reduced mobility (PRMs). All interviewees were asked for consent and informed of the purpose of the interview. Most of the interviews were conducted by telephone. They were recorded and transcribed and transcripts were sent to the interviewees for review and approval. Additional information was provided by e-mail. The protocol for the interviews loosely follows Christmann (2009) who recommends 'a strongly structured guideline which is similar to a standardised questionnaire, the only difference being that it includes more open questions than the latter'.

The question sequence was based on an assumption that the sequence of activities would resemble the ordering of a design process described by the Engineering Design Centre, Cambridge (Clarkson et al. 2007). First, discover the users' needs; understand the needs; translate the needs into requirements; create concepts to fit the requirements, then develop solutions. Each step has feedback into the previous step, which reflects the iterative nature of the design process.

The first question was concerned about planned design processes. Subsequent questions were related to how and where users were involved in each step of the design process. The questionnaire is available at inclusivelydesigned.wordpress.com.

5 Results

5.1 Light Rail Operator Scandinavia

The industrial design co-ordinator was interviewed about a new section of an urban rail system. The design process is embodied in a master document which formalises the translation, concept development and validation steps through repeated feedback to interested parties including users. The process is an evolution of one used for an earlier stage of the transit system. Formal documentation is generated for the elements (e.g. doors, seats, ticket machines, etc.), any problems and grounds for their resolution. This forms the cross-checking stage of the specification with the users' needs. Initial user-feedback led to a specification, described as a 'check list'. Initial ideation was by the operator and then by an outside agency. The operator's ideation used Post-It brainstorming following a year of study tours. Concept selection was done using a forum with representatives from disability organisations. The requirements translation did not seem to be explicitly user-centred nor were development and validation. Instead the documentation was used to refer back to the user-requirements.

5.2 Heavy Rail Operator Scandinavia

The subject was a former senior designer. The planned design process required concepts to be created in-house and then developed by an external consultancy. There was a strong stated emphasis on user-centred design. The firm used workshops, ethnographic observation, video-ethnography and consultation with representatives of disability interest groups. Ideation was not explicitly described. The intention for the project was that wherever possible, standard solutions were to be used (in theory obviating the need for ideation). Translation of the specification into requirements was done in consultation with interest groups. The users' role during concept development was not explicitly described. For development and validation the firm took a recently completed commuter train as a starting point; disabled-user interest groups were consulted on the design at the mock-up stage, during specification drafting and during model development.

5.3 Rolling Stock Builder 'A'

The subject was the chief vehicle architect. To incorporate the needs of PRMs, this multinational firm makes compliance with legislation and standards 'the key thing' but the organisation is 'very active in writing those standards in the first place'. Stakeholders are involved primarily before projects become active. The firm's design process consisted of a concept phase, a requirement description, a detailed concept phase, a development phase, which is 'preliminary design', and then a detailed design phase. Ideation occurs during the concept phase but no further description of what this entailed was forthcoming during the interview. The subject attributed the interior design to their 'internal design direction' and also to research and 'development in materials and manufacturing methodologies'. The translation of the specification into requirements is 'through standards and legislation'. Virtual modelling is used to test concepts with user-groups in the initial stage of the project but development did not focus on user groups' feedback. CAD models were preferred to physical ones for reasons of cost.

5.4 Rolling Stock Builder 'B'

The interview subject in this multinational firm referred chiefly to regulations and standards regarding elderly and disabled users' needs. Any additional requirements were provided by the operators. Concerning the planned design process and ideation stage, 'B' has 'no special design process for the elderly and disabled. It's part of the normal design'. The requirements are turned into finished designs using a process subsections handle specific areas. Translation of the specification, concept creation, development and validation were all described as following the 'normal design' process. Users could be accommodated in the design process if the customer requested it.

5.5 Design Consultancy 'X'

'X' is a French firm which designs high speed, regional and local trains. The interview subject was the design director. The planned design process was not fully described. The subject put a strong emphasis on direct observation in the field to see how users behaved, but the results focused on cultural differences rather than those related to age and disability. User data was also provided by the client. No data was provided on ideation or the translation of the specification into designs. When asked if users consulted about concept design, the subject responded that in PT, 'experience and proven solutions' and 'opinions from top management' mattered more. His opinion was that automotive and consumer product design had more rigorous and user-centred processes. Development and validation involved full-sized mock ups. The subject viewed the cost of these as being negligible in proportion to a contract's total cost.

5.6 Design Consultancy 'Y'

'Y' is based in the Midlands, UK, and carries out interior design for new urban light rail, for heavy rail and for renovations of heavy rail carriages in Europe and Asia. The interview subject was a senior designer. The response relating to the design process indicated that their custom was not to work to a prescribed model. When asked how the needs of PRMs were incorporated into the design process, the subject said there was 'no differentiation' between the needs of different groups. To find out about the users, 'Y' works with user groups and studies the user profiles created by their clients. Focus groups, interviews and personas were also deployed by the firm, which consults with the Royal National Institute for the Blind and an ergonomic specialist. Video ethnography was also cited, specifically with regard to parent–child groups, chiefly to work around the problem of communicating with younger children. The subject responded that the brief was where the most important work was done in terms of ideas. It was not made clear, if ideation was done as part of the creation of the brief or whether the brief only guided ideation. About whether users were involved in validating concepts, the subject tried to suggest that the firm's previous experience and the client's view were sufficient. No clear information was yielded concerning translation. Partial full-size models were made but no information was provided on user-testing of these models.

5.7 Design Consultancy 'Z'

This London-based firm works in transport, product and packaging design. The interview subject was the firm's director. The firm has a planned design process: 'We adopt the same approach for each of our projects, regardless of the specific industry…a framework of development milestones…'. Asked whether the process corresponded to the waterfall model of ID, the answer was 'Broadly speaking, yes….Every project follows rigorous steps of research, understanding, technical requirements, fine tuning and testing et cetera…'. For the EU market, 'Z' uses a specialist ergonomist to incorporate the needs of PRMs into the design process. This specialist was 'part of the project team'. The firm also invites disabled persons to address designers on their travel needs. The firm also works 'with partners who specialise in qualitative research, ethnographic observation and semiotics' and they 'observe in situ'. Modelling is done using CAD and hard-models. Full-size mock-ups are used to test users' responses and the ergonomic aspects. This modelling stage merges with the validation process in that resolving the models brings problems to light which are then corrected on screen before final 1:1 validation.

5.8 Design Consultancy 'C'

'C' is a small consultancy run by the subject. The firm has worked for several decades with a regional PT provider in southern Germany. The planned design process was described as 'systematic' starting with a problem analysis, definition of constraints and user investigations. Then follows ideation and consultation with the client (both transport provider and producer). The requirements are defined. Via CAD and hard model mock-up, the design proceeds to testing. The subject's response regarding the incorporation of PRMs into the design process referred to user interviews and to industrial standards. The user interviews focused on critical areas. Disability groups are consulted during the design process. Additionally, the subject cited his own experience of being more than 70 years old as informing his goals. Users are invited to test proposals when the design has reached the full-size, mock-up stage. Development and validation run semi-concurrently. The subject provided no information on ideation, translation or the role of users at the concept stage. The implication is that user involvement is loaded at the beginning and end phases of the process.

5.9 Design Consultancy 'D'

'D' is a large south German consultancy. Its clients are manufacturers and operators. The subject was a senior partner. The subject described the order in which the elements of the train are determined, from principal cross-section to local detail. Answering orally, the respondent alluded to a generic design process rather than a specific in-house ideal of such a process. The firm views the needs of PRMs as one of many parameters to be considered. Directly and indirectly, the subject suggested that it was through their client's relationships with special interest groups that their needs were accounted for: 'The first and best method is to learn from the experience of the operators of trains'. In two listed cases 'D' met with users in the existing trains and in mock-ups at the 1:1 modelling stage. The translation process did not typically involve users. If there was user involvement at this stage, it was through contact between the operator and user groups. The subject said that the firm did not typically consult users in between ideation and the building of mock-ups. Development and validation were merged. Users were consulted on specific matters, e.g. toilet layout and grab-handles.

6 Analysis of Results

Table 1 shows the 'user-involvement score' of each respondent. It notes the type of user-testing and the number of design methods used. Direct user-involvement is where users participate during development, i.e. communicate in person. Indirect involvement includes user-data provided by clients and input from standards and legislation.

The table was arrived at by counting instances where users were directly involved in the design process. It is not possible to guarantee that a higher user-involvement score means better design. It would be difficult to devise an objective weighting scale but there is some value being able to see instances of user-involvement. One can conclude that the mapping of PT design processes to the ID process is at best partial. Most of the reported methods and strategies that are cited are common to both mainstream industrial design and inclusive design. The main

Table 1 User-involvement, user-testing, methods

Name	Direct involvement	Indirect involvement	User testing	Methods
Light rail operator	5	1	1:1 model	5
Heavy rail operator	5	n/a	1:1 model	3
Builder A	2	6	CAD, detail	3
Builder B	0	2	No data	1
Consultancy X	2	3	1:1 model	2
Consultancy Y	2	2	No data	3
Consultancy Z	2	8	1:1 model	5
Consultancy C	8	9	1:1 model	2
Consultancy D	2	2	1:1 model	4

exception to this was Consultancy Z. Although the involvement of users and surrogates was not explicit in the description of the later stages of Z's process, it was implicit through the integration of experts in the design process, consultation with interest groups and the empathic methods used. In contrast, Consultancy Y said that '…inclusivity is such an important motivation behind what we do it isn't differentiated, it's part of the whole technics of the whole process…'. The interview however, did not seem to back up this claim to the extent the statement suggested. Builder A's methods were conventional. Builder B had low scores for all the parameters.

The Scandinavian light rail operator and heavy rail operator have similar scores with different design outcomes: the light rail system is flat floored throughout while the new heavy rail train is not. Both operators have higher direct-user involvement scores than either the design agencies or the producers. Qualitative differences moderate this description. For example, Consultancy Z scores 2 for indirect involvement, as does Manufacturer A. But for Consultancy Z the indirect user-involvement means having an internally placed advisor working on the project all the way through. For Manufacturer A, it means reference to the legislation and data provided by the client.

Consultancy C is interesting and ambiguous. The designer, as an older person, can directly empathise with other older people and their needs. However, he is one individual and not entirely representative. He does exemplify empathic design in that actually being an older user allows a better insight than when a 27-year old person role-plays the effects of ageing. The designer's involvement can be counted at each step of his design process but whether it is described as direct or indirect user involvement is open to debate. We have chosen to count it as indirect.

Direct and indirect user-involvement serve different ends. For example, Standards offer consistency; specialist advisors offer support to the client's unique design needs. Ideally, standards form a minimum level of user input but can't provide the rich information that direct involvement can.

7 Discussion

It has not been feasible to verify the claims made in the interviews. There exists also a gap between the claim that some user requirements have been met and that most or all user requirements have been met. It is not clear how respondents define meeting user requirements.

The processes reported here are primarily variants on a standard engineering model of design though the consultancies used a more complex process structure. Where user-centred techniques are deployed, it is at the start of the process and then, less often, at the end, during validation. The design consultancies demonstrated a broader use of inclusive design methods. The owner-operators had a higher level of user involvement than the manufacturers. Why have designers of PT generally not used in ID models? Perhaps, because models developed initially for product design reflect tighter consumer-product relationships. Another reason is institutional: the manufacturers are very large firms with a much greater division of labour. A written and oral interview of Rolling Stock Producer 'B' revealed very little of its design process other than that, by inference, it has a diffuse design process where no single individual has design authority. The case of the Scandinavian Light Rail operator is an example of how it might be done within a large company.

There were differences between the written and oral interview responses. In their verbal response, consultancy D avoided putting a special emphasis on design for disability: this was left to the discretion of the builder or operator. But the written answer reports that users are involved in consultation, concept approval and prototype-testing. Information on users' needs was found directly through the firm's own research and data provided by the client. This suggests that different responses might have been gained if all the subjects had provided full, written answers. One interpretation is that written responses may reflect the wider view of the corporation. Alternatively, perhaps the perceived formality of written answers might lead the respondent to give a more 'officially acceptable' result than would the informal verbal response. We suggest that the verbal answers, though less precise, might be more representative of the firm's actual modes and attitudes, even if they yielded less precise descriptions of activities.

Producer 'A' placed a strong emphasis on standards. The answers tended to avoid detailed reference to specific design activities or specific instances of user-involvement. The respondent oscillated between the views that it was either a difficult target to attain or that it was also an ordinary, routine requirement. The view that the standards, which they themselves helped to draft, are sufficient should be contrasted with Tyler (2002, cited in Wilson 2003) who wrote that it is very easy for a minimum standard of accessibility to become a norm, retarding innovation. Tyler recommends results-based standards in place of process standards or quantitative targets (minimum door widths, for example). This leads the discussion to the case of the Scandinavian heavy rail operator which focussed on deliverables but with mixed results. The operator's process listed only deliverables (not means). The goal of level access for wheel chair users was attained but the other requirement for diesel propulsion meant the rest of the carriage is inaccessible. User-driven design was not part of the tender which focused on descriptions of a single concept of a finished train at the lowest possible cost.

Operators who order custom rolling stock can ensure that users' needs are addressed but only to the extent that they are able to specify the details of equipment. Design consultancies' freedom to use inclusive design approaches is constrained by the customer's commitment. Manufacturers who design in-house are poorly positioned to address user needs since they lack the operators' customer insight or a habit of user-research found commonly in design consultancies.

In some cases, and for valid reasons, maintaining the user's presence in a long, complex process is more difficult than in consumer product design. In which case, the standard model of inclusive design might turn out to be less well suited to design for rail transport than it is for product design and assistive technology.

References

Azmin-Fouladi N (2007) Designing the inclusive journey environment. In: Proceedings of include 2007, Helen Hamlyn Centre, London

Barnes C (1991) Disabled people in Britain and discrimination. Hurst and Company, London

Baird L, Dong H (2009) Inclusive design for air travel. In: Proceedings of the include 2009, Helen Hamlyn Centre, London

British Standards Institute (2005) British Standard – BS 7000-6:2005. Guide to managing inclusive design

Christmann G (2009) Expert interview on the telephone: A difficult undertaking. In: Bogner A, Littig B, Menz W (eds) (2009) Interviewing experts. Palgrave Macmillan, Basingstoke, UK

Clarkson PJ, Coleman R, Hosking I, Waller S (2007) Inclusive design toolkit. Cambridge Engineering Design Centre. University of Cambridge, UK

Davis G, Mills A (2005) A common system of passenger safety signage. In: Wilson JR, Norris BJ, Clarke T (eds) Mills A Rail human factors: Supporting the integrated railway. Ashgate Publishing, London

Evans G (2005) Accessibility and user needs in transport design. In: Proceedings of the include 2005, Helen Hamlyn Research Centre, London

Hirsch L, Thompson K (2011) Tarzan travellers: Australian rail passenger perspectives of the design of handholds in carriages. In: Proceedings of the HFESA 47th annual conference. Ergonomics Australia – Special Edition, 11(34):1–5

HM Government UK (2010) Equality Act. Available at: www.legislation.gov.uk/ukpga/2010/15/part/1 (Accessed on 4 November 2013)

Kaneko S, Hirai Y, Elokla N (2009) Bus for everybody: Nishitetsu Bus case study in Japan. In: Proceedings of the include 2009, Helen Hamlyn Centre, London

Kitchin R (1998) Out of place, knowing one's place: Space, power and the exclusion of disabled people. Disabil Soc 13(3):343–356

Langdon PM, Waller S, Clarkson PJ (2009) Just get a ticket: Inclusive design in action. In: Proceedings of the include 2009, Helen Hamlyn Centre, London

Lauren J, Rhind DJA, Robinson K (2005) Rail passenger perception of risk and safety and priorities for improvement. In: Wilson JR, Norris B, Clarke T (eds) (2005) Rail human factors – Supporting the integrated railways. Ashgate, Aldershot

Mackie E (2012) Design for public transport. In: Tovey M (ed) Design for transport. Gower, Aldershot, UK

Metz D (2003) Transport policy for an ageing population. Transp Rev A Trans Transdisciplinary J 23(4)

Marshall R, Porter JM, Sims R, Summerskill S, Gyi D et al (2009) The HADRIAN approach to accessible transport. Work A J Prev Assess Rehabil 33(3):335–344

Miller P, Gillinson S, Huber J (2006) Disablist Britain. Demos, London

Nickpour F, Jordan PW, Dong H (2012) Inclusive bus travel – A psychosocial approach. In: Langdon PM, Clarkson PJ, Robinson P, Lazar J, Heylighen A (eds) Designing inclusive systems. Springer, London

Social Exclusion Unit (2003) Making the connections: Transport and social exclusion interim findings from the Social Exclusion Unit. HM Stationery Office, London

Tennoy A, Lieren M (2008) Accessible public transport: A view of Europe today, policy, laws and guidelines. Report 952/2008. Institute of Transport Economics, Oslo, Norway

Tyler N (ed) (2002) Accessibility and the bus system: From concepts to practice. Thomas Telford Publishing, London

Wilson JR, Norris BJ (2004) Rail human factors, past, present and future. Appl Ergon 36(6):649–660

Wilson JR, Norris B, Clarke T (eds) (2005) Rail human factors – Supporting the integrated railways. Ashgate, Aldershot

Wilson M (2003) An overview of the literature on disability and transport. Disability Rights Commission, London

Woodcock A (2012) User-centred transport design and user needs. In: Tovey M (ed) Design for transport. Gower, Aldershot, UK

Inclusive Strategies for Universal Access in Educational Campus Environments

G. Raheja and S. Suryawanshi

1 Introduction

Education is a key to employment opportunities, higher productivity and income which leads to better health and social participation. It also helps in enhancing individuals' capabilities for self-sufficiency (University Grants Commission 2009). Higher education has its own importance in terms of employability, political power and national development. In the Indian context, higher education scenarios lack inclusive approaches and therefore exclude persons with disabilities. A nation making efforts to join the league of developed nations cannot afford to move ahead without including the weaker members of society. With about 10 % of the population with disabilities and a huge dropout rate of students with disabilities from early schools, there is an alarm call for the nation to devise inclusive measures and strategies to make education accessible to all. The universal design approach becomes a significant way to move ahead since the Enforcement of the Persons with Disabilities Act (1995), and having ratified the United Nations Convention for Rights of Persons with Disabilities (United Nations 2006) in the year 2007, India is committed to an inclusive vision for persons with disabilities in education and all other sectors.

> Higher education being at the apex of the educational system is an essential input for meeting the manpower requirements of the highest caliber in the crucial areas of national development. It is also an important contributory factor for ensuring social justice by providing vertical mobility to deprived sections of society by making higher levels of knowledge accessible to them and, in the process, improving quality of life of the nation as a whole.
>
> (Azad 2008)

G. Raheja (✉) · S. Suryawanshi
Department of Architecture and Planning, Indian Institute of Technology
Roorkee, Roorkee, India
e-mail: gr.iitr@gmail.com

P. M. Langdon et al. (eds.), *Inclusive Designing*, DOI: 10.1007/978-3-319-05095-9_15,   165
© Springer International Publishing Switzerland 2014

As increasing awareness about the importance of education and the schemes instituted by government increases most of the educational campuses in India are addressing diversity in students.

2 Inclusive Design in Educational Campuses

Traditionally special education systems were developed to educate children and youth with disabilities, which excludes them from mainstream education. A special education system does not make children and youth with disabilities part of society able to participate without discrimination. The Salamanca statement passed by UNESCO (1994) supports the practice of inclusive education for students with disabilities with the caution that

> while inclusive schools provide a favourable setting for achieving equal opportunity and full participation, their success requires a concerted effort, not only by teachers and school staff, but also by peers, parents, families and volunteers.

(Ainscow 1994)

Inclusion in education refers to fulltime integration and accommodation of all the students with diverse abilities as well as providing systems of support for all activities. Inclusive global education, a new concept introduced, defined and highlighted by Landorf and Nevin (2007), is a way to honour the diverse cultural, linguistic, physical, mental and cognitive complexities of students. Gomes (2008) suggests the importance of equitable and inclusive education for all based on research with multicultural and multiracial Canadians. Whereas Peters (2003, 2004) in World Bank reports highlight the outcomes of inclusive education in building one's self-esteem in social life through a positive attitude towards learning. This can be increased by the participation of parents and communities.

Perceptions about inclusive education differ contextually in developed and developing countries at the global level. In Africa, students with disabilities still struggle to get access to education due to lack of meaningful implementation of inclusive education policies (Chataika et al. 2012). In India, Singal (2009) states, the meaning of inclusive education is taken as inclusion of marginalised groups with more emphasis on religious groups, girls and categories like scheduled castes and scheduled tribes than on disabled students. Implementation of universal design in education details various aspects of the educational system which need to be designed for all. These various aspects range from entry or admission to campus to virtual education on the campus. Universal design in education provides a philosophical framework for designing a broad range of educational products and environments (Burgstahler and Cory 2008).

3 Significance of Study

This chapter while ideating the concepts of inclusive design approaches for educational campuses is based upon a live research study done with students with disabilities and other stakeholders on the campus of a prestigious national university, Jawaharlal Nehru University (JNU), New Delhi. The study is based upon the initiative taken by the campus administration along with disability groups and other stakeholders to facilitate universal access and develop a strategy for inclusive education on this higher education campus. As a result, it would not only benefit the campus community but would become a role model for other national universities and campuses to devise and adopt similar strategies to make them inclusive.

Educational campuses act as a small city within a city, where people of varied culture, gender, caste, size and ability work together. User mapping was employed as a strategy to profile stakeholders in the educational framework from the perspective of access (Fig. 1).

Physiogenic and sociogenic factors affect the self-development of youth in higher education. For youth with disabilities, these factors should promote inclusion rather than exclusion at different levels of studies. Most nations have laws and policies on inclusive education as well as building standards to provide an accessible built environment. Implementation and monitoring of these on an educational campus where people with diverse abilities and varied culture live is still an under-researched domain especially in India. Universal design helps to make new educational campuses inclusive to the widest range of individuals by eliminating the need for retrofitting or reconstruction of any built environment. However, retrofitting solutions remain a challenge to implement in most contexts where the built environment already exists. To apply universal design in education diverse elements (like parking, entrance, door handles, widths, circulation, toilets, illumination, emergency exits, etc.) need to be considered to provide maximum usability for all.

Students remain the major users of all facilities and services provided on educational campuses. Students in higher education, especially in national universities, form diverse population groups on campus, ranging from social to cultural, rural to urban, able bodied to persons with special needs backgrounds. Application of inclusive educational strategies in all services related to students would help remove barriers to access and ultimately benefits all other users on campus.

3.1 Selection of Case Studies

The case study of JNU was selected by the authors to showcase an understanding through a firsthand experience of access auditing this national university. Established in the year 1969, JNU is an epitome of higher education. Spread over a

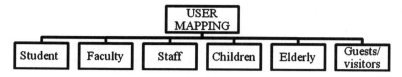

Fig. 1 User mapping framework for educational campus

campus of over 1,000 acres, on the Aravali Range, JNU houses over 7,300 students with over 500 faculty members along with other staff members. The contoured landform of this selected campus supporting its built infrastructure is in itself a uniquel site which has diverse consequences for the people who use it daily. Persons with disabilities and those with limited adaptive capabilities negotiate the barriers of the built and natural environments of JNU campus on a regular basis.

4 Research Design and Methodology

This chapter uses a field-based case study approach to document, understand and share lived experiences of diagnosing, auditing and developing retrofitting solutions for educational campus environments. Lessons learnt thereby guide a holistic approach to inclusive strategies for implementation. The qualitative study is conducted with the help of students with disabilities on the JNU campus. Participatory research remains an integral approach for this study to arrive at a broad understanding (Iwarsson and Stahl 2003; Sharma 2012; Ambati and Ambati 2013). It helps in recognising the importance of an accessible environment for the different needs of the diverse population of students on the campus through their experiences. This study attempts to find out physical barriers for the students with diverse abilities on the campus and its impact on their educational experiences.

This exploratory research was conducted through five stages and using multiple tools for data collection, as shown in Table 1 on the next page.

4.1 Sampling

This research used chain-referral sampling, a synonym for snowball sampling that is often used to find hidden populations. Identified group of students with disabilities were not easily discoverable on site. Of the total population of 7,304 students at JNU, about 187 were grouped in the category of students with disabilities (JNU 2012). Through the social networking of the student team the authors could contact them and discuss various issues related to physical accessibility on the campus. As the time available to conduct case studies was less, a total of 89 students with disabilities of both genders along with one faculty member were

Table 1 Table showing different stages of work with multiple tools used for each stage

Stage	Identification	Description
1	Reconnaissance site survey	Visual survey Data collection through site plans Focus group discussions
2	Plan annotations	Building drawings Physical survey of each building Marking of physical barriers observed on drawings Annotating drawings with measures for improvement of accessibility
3	Development of access audit tool	Analysis of stage 1 and stage 2 Analysis of focus group discussions The literature review Formulation of access audit tool considering as many building elements as possible
4	Pilot study and validation	Selection of three buildings with varied activities Access audit of these buildings with designed tool Analysis of stage 3 and stage 4 Study of outcomes of pilot access audit Modifications to audit tool
5	Detailed access audit	Access audit in detail for all selected buildings Plan annotations Observations in form of still photographs, video clippings. Physical measurements of building elements. Simulating special requirements experiences by self (Khare and Khare 2012) Documentation of all barriers in built and un-built campus environment (connecting pathways between buildings)
6	Design recommendations	Analysis of all above stages General design recommendations as per building elements Some of the common recommendations applicable for all buildings Specific design solutions like accessible toilet for residential block

approached. In the enrolment for various programmes offered by JNU, the number of students with vision impairment and with physical disabilities was found to be high. Physically disabled students, using a variety of assistive aids like crutches, manual or electric wheelchair, three-wheeled scooters, tricycles, white canes, hearing aids, etc. were selected in the sampling population to maximise the range of special requirements surveyed to better improve usability of the campus.

5 Results and Discussions

While the study produced detailed series of data collection and reviews, this chapter highlights the results and discussions based upon an academic typology of campus building. In it, access audits of a typical academic building at JNU campus have been discussed in detail.

5.1 Perspectives of Students with Disabilities

Focus group discussions were carried out with students with diverse disabilities. The discussions produced information about different factors that impede physical access in the educational built environment.

One of the vision-impaired students shared their difficulty in navigating with the help of a guide cane, as path navigation blocks were absent in academic as well as residential built spaces. A student with lower limb dysfunction, who crawls using hand support, along with other vision-impaired students emphasised the problems of undulating sites with unpaved surfaces near the students facilities and also issues of horizontal and vertical circulation. Inside the residential hostel, too many changes of level between residence and dining blocks and the absence of navigating guiding blocks and ramps-forced students with disabilities to take inconvenient longer routes to reach their common dining spaces. These inter-connecting routes also have step levels coupled with poor maintenance. Some of these students said that they suffer high fatigue levels as a result of awkward access routes which also hamper their studies.

One of the students with a physical disability notes the inaccessibility of the multipurpose activity room located on the first floor of the hostel because of the absence of lifts in the three story building. Most of the students with disabilities faced problems of inaccessible toilets (w.c. and bath inclusive) which emphasized the immediate need to create an accessible toilet block on the ground floor at least. A female student with vision impairment described from her own experience of using the residential block, the potential barriers of level differences between building blocks, temporary obstructions in passages like shoe racks, fire extinguishers, hanging electric wires, etc. Another female student with low vision explained the problems caused by absence of railing in the balconies of ground floor rooms. With level difference of approximately one to two feet between balcony and open landscape space in front, it makes them vulnerable.

The Helen Keller Centre in the main library building with facilities like Jaws and other screen reading softwares, Braille printer and sensitised staff members were pointed out as a unique support facility for the vision impaired students on the JNU campus. At the same time, some students emphasised the condition of pathways leading to the library and other schools within the campus which make persons with vision impairments vulnerable navigating their way through and cause discomfort to other mobility impaired students. Students using wheelchairs mentioned the problem of inadequate width of lift opening and tight manoeuvering space inside which made it hard for them to approach upper floors in academic buildings. This forces them to have a carer along with them and increasing their dependence. Door width(s) of faculty rooms and their internal furniture arrangement were also seen as a barrier by a few non-ambulant students using wheelchairs. Addressing the above-mentioned issues would definitely be a move towards inclusive education.

5.2 Perspectives of Administrative Authorities

Access for all and inclusion in education do not happen by accident but every implementation needs administrative will and intent. In a historic decision, in support of the implementation of Persons with Disabilities Act (1995), the JNU administration formed a committee headed by the Vice Chancellor to look into the needs of persons with disabilities on campus. The Helen Keller Unit in the library was one of the early initiatives taken by the administration to provide education support for persons with vision impairments. Being a socially vibrant and active campus, JNU administration took on board the student representatives of diverse disabilities and other student leaders in the process of decision making. The authors as part of the access consultation team were appointed to suggest design recommendations that could be implemented to initiate steps towards inclusion. It is important to note that the administration aspired to establish this campus as a role model of universal access and inclusive design to showcase a successful model of inclusion at an international level.

6 Access Audit Observations

Access audit(s) of independent school buildings on campus, hostel residential spaces and the common amenities area were taken up by the author(s) as an access consultancy team. Based on the focus group discussions with students and faculty members with disabilities, administrative authorities, physical site observations (measurement based and visual) and simulating personal experiences, these audits were carried out on JNU campus.

As an illustrated example from a set of 25 original built structures, access audit observations of the School of Social Sciences at JNU campus are discussed below as a typical case, where students from national and international communities seek education at various academic levels. Every academic year a total of approximately 250 students with diverse backgrounds get admitted under this discipline at postgraduate and doctoral levels. Observations on the subject of major physical barriers and their retrofitting solutions are discussed under the following headings:

6.1 Parking and Entrance

Reserved parking for persons with disability was not available on site as recommended per as norms. Allocated parking for two wheelers and four wheelers was recommended at the nearest point to the entrance lobby. The ramps provided were steep with a gradient of 1:12. Also, the existing handrail details were not as per the accessibility guidelines and an additional ramp with gradient of 1:12 with path navigating blocks and handrails as per guidelines was suggested.

6.2 *Horizontal and Vertical Circulation*

Absence of path navigating blocks throughout the building makes things difficult for students with partial and complete vision impairments. Provision of path navigating blocks in all the circulation areas including toilets was suggested by drawings. In the vertical circulation of the building, staircases are provided at front and back. The suggestion of handrails on both sides extending beyond the last step, use of colour contrast for risers, a curb rail at the lower level of railings, closing of space below soffits will help all vision-impaired students. The suggestion of keeping circulation corridors free from any obstructions like fire extinguishers was most meaningful for vision-impaired students as they always face the problem of hitting their heads, while moving through corridors.

6.3 *Services*

The provision of accessible toilet(s) on each floor of the building was accepted as a proposal. Solutions for the problem of manoeuvering spaces inside toilets, height of the wash hand basin and level of latch were developed and discussed with the architect of the building. At the drinking water fountain, one lower water fountain with clear knee space was suggested to meet the needs of students with wheelchairs.

6.4 *Building Interiors*

Provision of ramps and handrails on both side walls of the seminar hall will help the student to reach all seats provided. Height of door latch from finished floor in faculty rooms, provision of vision panels to the doors of classrooms, increased door width in some faculty rooms were some other suggestions to meet the needs of students with physical disabilities.

6.5 *Signage and Wayfinding*

Signage and wayfinding remains an ignored dimension of most educational campuses in India. Signage for the low-vision and hearing-impaired students was recommended with high contrast values and highlighted to ease the navigation process from entry to exit as well as internal circulation. This would help in navigating with independence.

In the stakeholders review meeting, administrative authorities discussed the implementation of all the solutions to provide maximum accessibility for all students with diverse needs. The process of review was done at two intermediate stages, one at sketch draft and another at final level.

7 Inclusive Planning Process

The planning process and evolution of access strategies adopted a three-step approach in this project, viz. Visual identification and documentation of barriers, participatory discussions with stakeholders (users with disabilities, administrative authorities, site implementation teams) and consensus building through drawing feasibility presentations of access solutions. Implementation measures were recommended in a holistic accessibility master plan by building an overall rationale rather than by ad hoc site decision-making. Democratic participation by stakeholders made the whole approach inclusive in nature. Since the implementation is guided by the availability of funds, it was consensually accepted to prioritise need based on numbers of stakeholders with each disability, the severity of barriers in the structures and the nature of the site. Through careful observations, technical discussions and feasibility iterations, design solutions to improve accessibility were developed with keeping stakeholders involved in the overall process. Recommendations in text and drawing formats for easy implementation were provided by the authors. Finally, solutions for all 25 buildings were developed, discussed and presented for review to stakeholders before initiating the implementation process of an inclusive campus. Building standards as prescribed in the National Building Code (Bureau of Indian Standards 2005) and Accessibility Standards as issued by Ministry of Social Justice and Empowerment, New Delhi were used as reference documents to suggest accessibility measures for implementation to ensure inclusivity in space.

8 Conclusions

This study highlights a comprehensive process for creating universal access on educational campuses. With a case-based explanation and discussion on JNU campus, it emphasises a participatory approach that was implemented in order to map the access needs of persons with disabilities with reference to their educational environments and their experiences within them. Through a field-based survey approach, focus group discussions, visual documentation techniques and self-simulation experiences, this research showcases a methodology for inclusive planning and design for implementation in campus-built environments. Towards the end, this chapter highlights the perspectives of students with disabilities along with an administrative view on the issue of universalising access. Further, it illustrates the access audit observations of a typical school building on campus. It is envisaged that most campuses of higher education or otherwise could adopt similar processes to initiate inclusive planning at grass roots level.

References

Ambati NR, Ambati H (2013) Paradigm shift in German disability policy and its impact on students with disabilities in higher education. Int J Soc Sci Interdis Res 2(4):22–42

Azad JL (2008) Financing and management of higher education in India. Gyan Publishing House, New Delhi

Bureau of Indian Standards (2005) National Building Code-2005, Government of India

Burgstahler S, Cory R (eds) (2008) Universal design of higher education: from principles to practice. Harvard Education Press, Boston

Chataika T, Mckenzie JA, Swart E, Lyner-Cleophas M (2012) Access to education in Africa: responding to the United Nations convention on the rights of persons with disabilities. Disabil Soc 27(3):385–398

Gomes M (2008) Creating an equitable and inclusive institution. Discussion paper. Available at: www.viu.ca/humanrights/docs/CreatinganEquitableandInclusiveInstitution.pdf. Accessed on 4 Nov 2013

Iwarsson S, Stahl A (2003) Accessibility, usability and universal design—Positioning and definition of concepts describing person-environment relationships. Disabil Rehabil 25(2):57–66

Jawaharlal Nehru University (JNU) (2012) Annual report of 2011–12. Jawaharlal Nehru University, India

Khare R, Khare A (2012) Teaching universal design through student design competition. Spandrel 4:104–114

Landorf H, Nevin A (2007) Inclusive global education: implications for social justice. J Educ Adm 45(6):711–723

Peters SJ (2003) Inclusive education: achieving education for all by including those with disabilities and special education needs. Paper prepared for the Disability Group, World Bank, Washington

Peters SJ (2004) Inclusive education: an EFA strategy for all children. World Bank, Washington

Sharma A (2012) Higher education and its perspectives with special reference to differently able learners. Int Multidis E-J 1(6)

Singal N (2009) Education of children with disabilities in India. Paper commissioned for the EFA global monitoring report 2010, reaching the marginalized

The Persons with Disabilities Act (1995) The Persons with disabilities (equal opportunities, protection of rights and full participation) act, 1995, part II, section 1 of the extraordinary gazette of India. Ministry of Social Justice and Empowerment, Government of India

UNESCO (1994) The Salamanca statement on principles, policy and practice in special needs education. In: World conference on special needs education: access and quality. Salamanca, Spain

United Nations (2006) Convention on the rights of persons with disabilities. Available at: http://www.un.org/disabilities/convention/conventionfull.shtml. Accessed on 19 Nov 2013

University Grants Commission (2009) Annual report 2008–09. University Grants Commission, India

How do People with Autism (Like to) Live?

M. Kinnaer, S. Baumers and A. Heylighen

Abstract Research on inclusive design focuses on designing environments that account for the diversity in human abilities and conditions. People with autism, for instance, deal with their environment in a particular way because their different way of processing information influences their spatial experience. Literature offers a growing number of concepts to create autism-friendly living environments. These concepts start from putting people with autism centre stage, yet in their formulation the autistic person him/herself often risks to disappear from view. This raises the question what meaning and value these concepts have, and how designers can use them. The study reported here aims to reconsider these concepts by refocusing on autistic people themselves. Interviews were conducted with 11 adults with autism who are living more or less independently and were willing to share their stories about how they (like to) live. On the one hand, analysis of these interviews shows that concepts of autism-friendly architecture are not indisputable rules that can be applied straightforwardly, and that one concept may reinforce but also counteract another. In each particular situation thus a balance must be sought, which will likely be easier when designing an environment for a single known inhabitant than when designing for multiple known or potentially unknown inhabitants. On the other hand, visits to autistic peoples houses often gave a sobering impression: very common houses where only details suggest that someone with autism is living there. Often, however, reality often does not reflect the ideal situation they described. The latter starts not so much from how it should be, but from how they would like it most, which does not necessarily fit the traditional view of a good place to live. As a result, this study contributes not only

M. Kinnaer · S. Baumers
Department of Architecture, University of Leuven (KU Leuven), Leuven, Belgium

A. Heylighen (✉)
Department of Architecture, Research[x]Design, University of Leuven (KU Leuven),
Kasteelpark Arenberg 1/2431, Leuven, Belgium
e-mail: ann.heylighen@asro.kuleuven.be

P. M. Langdon et al. (eds.), *Inclusive Designing*, DOI: 10.1007/978-3-319-05095-9_16,
© Springer International Publishing Switzerland 2014

to a more nuanced understanding of concepts of autism-friendly architecture found in literature, but also to a more colourful image of what an autism-friendly living environment could be.

1 Introduction

Research on inclusive design focuses on designing environments that respect the diversity of human abilities and conditions. Early studies concentrated on physical difficulties, and paid little attention to mental or cognitive conditions (Mostafa 2007), which however raise similar questions. People with autism, for instance, deal with the environment in a particular way because their different way of processing information influences their spatial experience (Baumers and Heylighen 2010). For this reason, researchers and architects started developing a vision of autism-friendly architecture, resulting in a growing number of concepts for creating autism-friendly living environments. While these concepts are meant to put people with autism centre stage, through the process of formulation and generalisation the autistic person him/herself risks once again disappearing from view. This raises questions as to what meaning and value these concepts have, and how designers can use them. So our study reconsiders these concepts by refocusing on autistic people themselves and discussing with them their (ideal) living environment.

The house is described as an individual's most important anchor in the environment, a place offering shelter, privacy, safety, control and status (Coolen 2006). It is said to be that unique spot in the world where one feels at ease and is fully in control. But is this also true for independent adults with autism? How do they live, and are they satisfied with their situation? What could be improved? What would their ideal place to live look like, and how does it relate to concepts of autism-friendly architecture? Our study addresses these questions based on interviews with more or less independently living autistic adults who were willing to share their stories about how they (would like to) live.

2 Autism-Friendly Architecture in a Nutshell

Concepts found in literature on autism-friendly architecture often relate to sensory or mental accessibility (Table 1). The former responds to sensory difficulties autistic people may face for each of the senses: hypersensitivity, hyposensitivity (Bogdashina 2003) and interference (Grandin 1995). This variety of sensory challenges inspired the advice to create a neutral environment (Brand 2011). To compensate for a lack or overload of sensory stimuli, two extra concepts are advanced: *sensory rooms*, offering rich multisensory experiences through textures,

Table 1 Selected concepts of autism-friendly architecture [henceforth referred to by]

Sensory accessibility	Mental accessibility
Sensory room [S1]	*Predictability* [M1]:
Escape space/privacy [S2]	*transparency* [M1a], *overview* [M1b]
Clarity and order [S3]	*Consistency* [M2]
Colours and patterns [S4]	*Comprehensibility* [M3]
Natural daylight [S5]	*Controllability* [M4]:
	personal space [M4a], *exits* [M4b]
Sounds from outside [S6]	
Enclosure [S7]	

light, colours, sounds etc., and *escape spaces*, providing privacy and safety to retreat from too demanding situations (Brand 2011; Vogel 2008).

Each of the senses serves as the basis for more specific guidelines. Difficulties in processing visual stimuli may increase autistic people's susceptibility to *colours*, visual *patterns* or *artificial lighting* (Grandin 1995; Bogdashina 2003). In response to this, researchers recommend environments that display a great sense of *clarity and order* (Brand 2011), e.g. by providing adequate (built-in) storage space (Vogel 2008). People with autism can also be either attracted or extremely sensitive to certain (mechanical) sounds, or have difficulties distinguishing foreground from background noise. Therefore, acoustic qualities should well exceed the existing minimum standards regarding both *sounds* coming *from outside* and reverberation (Mostafa 2010). Finally, when the other senses fail to give an accurate representation of reality, the way things feel becomes even more important (Grandin 1995). Therefore materials and textures are well-debated in relation to autism, as is the feeling of *enclosure*. Mostafa (2010), for instance, recommends providing small intimate spaces for quiet activities.

The attention to mental accessibility stems from the observation that for autistic people, reality can be terribly chaotic and unpredictable. Their struggle to imagine and anticipate certain situations requires environments to be *predictable* (Sanchez et al. 2011), e.g. through clear sight lines or *transparency* (Ahrentzen and Steele 2009; Brand 2011) offering *overview*. Besides structure in space, autistic people are said to require a lot of structure in time (e.g. through routines and schedules). Mostafa (2010) therefore stresses the importance of *consistency*, making sure that there is a place for everything and that everything takes place on the right spot. Predictability can also be enhanced by *comprehensibility*: Consistent spaces with a clear structure and evident clues may offer the necessary support in delineating, organising and using different spaces (Sanchez et al. 2011). These concepts relate to the need for *controllability*: a familiar, predictable environment can offer a hold and thus increase emotional security and the feeling of control. In this respect, *personal space*, psychologically as well as physically, can become highly important (Ahrentzen and Steele 2009). Communal rooms should be designed with ample spaces, taking into account that autistic persons' feeling of being crowded might differ from the designers'. When situations become too overwhelming, the design should allow for retreat to a more sheltered space, e.g. a subtle alcove or even a completely separated (escape) space (Vogel 2008). To increase the feeling of controllability, Ahrentzen and Steele (2009) recommend providing enough *exits*.

3 Methods and Participants

The way a (home) environment is negotiated, experienced and interpreted, several authors point out, cannot be accurately measured, quantified or grasped by straightforward logic (e.g. Bollnow 2011). As we want to understand how autistic adults (would like to) live, and how this relates to concepts for creating autism-friendly architecture, we thus adopted a qualitative and interpretive research approach. Adults with autism were invited to share their story through a call on the website of an autism knowledge centre, a self-help group for and by autistic adults, and snowball sampling. Five persons agreed to participate in an in-depth inter-view—four face-to-face, one through e-mail—where possible combined with a home visit. In addition, six persons participated in a focus group interview.

All 11 participants (five women, six men), whose name has been changed for privacy reasons, are more or less independently living adults with (almost) normal intelligence. John and Ken both live with their parents, Pat with her family; Mike is in a living-apart-together-relationship and all others live on their own, except for Maggie and Simon whose children sometimes visit. All live in a single-family home, be it a detached or semi-detached house or an apartment.

Interviews were semi-structured. The interviewer (the first author) introduced issues people confront when searching for a place to live, yet went more deeply into aspects participants brought up. Themes were chosen not to probe concepts of autism-friendly architecture directly, but to uncover which aspects play a role indirectly. Therefore the interviewer introduced a selection of concepts in the form of thematic dilemmas (Table 2), (Other dilemmas could be formulated based on concepts related to indoor climate, incidence of light, technical installations, etc., yet exceed the scope of this study). By presenting two alternatives, she advanced the concepts not as one solution to be confirmed, but as a trigger to start off the conversation. The concept of *consistency* [M2] (Mostafa 2007), for instance, was introduced by contrasting separate rooms with a loft. *Clarity and order* [S3] (Vogel 2008) was contrasted with the possibility of displaying everything.

All interviews were transcribed verbatim and analysed by assigning codes and categories on varying levels, using both concepts from literature and aspects arising from the interview itself. The next section reports on the aspects brought up by the participants in response to the thematic dilemmas, and relates them to the concepts (see Table 3). Translations of quotes into English are the authors'.

Table 2 Themes and dilemmas introduced during the interviews

Themes	Dilemmas
Neighbourhood	Quiet environment versus reachability
Housing type	(Detached) house versus apartment
	Big garden all around versus view over the city
Layout	Separate rooms versus loft
Interior	Everything stowed away versus everything displayed

Table 3 Concepts and aspects brought up in discussing themes and dilemmas (Aspects arising from the interviews themselves are marked with an asterisk)

Themes	Dilemmas	
Neighbourhood	Quiet environment: *escape space, sounds from outside, personal space*	Reachability: *independence**
Housing type	(Detached) house: *sounds from outside, controllability—exits*	Apartment: *privacy*
	Big garden all around: *sensory room, privacy*	View over the city: *predictability— overview, independence**
Layout	Separate rooms: *escape space/privacy, clarity and order, enclosure, consistency, controllability— personal space*	Loft: *natural light; predictability— overview, controllability— personal space*
Interior	Everything stowed away: *clarity and order, colours and patterns*	Everything displayed: *predictability— overview, consistency, comprehensibility*

4 Findings

4.1 Neighbourhood

For most people, choosing a suitable neighbourhood is a first important step when searching for a place to live. As Mike remarks: 'Put a perfectly adapted house in a wrong neighbourhood and it becomes a bad house'. Several factors come into play here. First, traffic appears often to be a disturbing factor because of the bustle and noise it causes [S6]. Ken dislikes living along a busy road, and Maggie complains that the noise in her street 'makes [her] crazy' [S2, S6]. Yet some participants live in a place others would avoid, precisely because of the nuisance. Pat lives next to the railway, but the trains' sound does not disturb her that much as she got used to it as a child. Also, John lives very quietly, he says, while only a few trees separate his garden from the highway. Apparently one can get used to certain sounds. Second, human traffic may cause bustle too. Tina likes the area around the central square in a nearby provincial town to go out, but in the evening there are so many people there that she could never live there. Maggie minds the kind of people living in her neighbourhood. She lives in 'that poor part of Brussels', where many drug addicts hang around and waste is disposed of illegally. She liked the area where she used to live better because many older people were living there. Mike also appreciates his building being occupied mainly by older people, be it because their social expectations are limited, while Daniel does everything to make sure nobody sees him, so that in the evening he can be quietly alone [S2]. Norah by contrast lives in a quiet area with many young families, and enjoys the activity of children biking. Once in a while she goes to the small square in front of her house to talk with her neighbours, which illustrates that the need for quietness [S6] and personal space [M4a] does not equal 'autistic people like to be alone'.

This need is further put into a perspective when confronted with reachability. Most participants live relatively quietly [S6], yet often have to seek a compromise with reachability, as evidenced by statements like 'a quiet neighbourhood for being so close to the centre'. Elaine lives close to a department store parking. Between 6 and 7 pm, when all cars are leaving, it is quite busy, but afterwards it is quiet again. A location close to the centre, at walking or biking distance from facilities, is often put forward as an asset. Yet, what should be reachable and how? For Pat, who can drive a car, the vicinity of her son's school is especially important, as are the connections to where she works. Tina, by contrast, needs public transport to reach everything, including her work, as she cannot drive. She takes evening classes, has a busy social life and likes to go out. When no public transport is available, she has to get home on foot or by bike. John has the ideal combination of quietness and reachability, he says, although for most facilities he needs to go to the next village. He likes living quietly in the country. The only thing scaring him somewhat is his own reachability. Because where he lives is very isolated, he would need to stand waving 300 m from his house for the emergency services to find him. Simon used to live in a caravan in the woods, but moved because the confrontation with loneliness and raw nature aroused many fears in him. A quiet location often comes with a considerable dependence on others, Mike notes, which may become a problem too.

4.2 Housing Type

Another dilemma the interviewer introduced regards the housing type: a (detached) house or an apartment? Besides less space, participants associate apartments spontaneously with nuisance from neighbours, above and next-door [S6]. If Maggie's neighbours quarrel or slam a door, the walls seem to be made of cardboard. Also in David's semi-detached house this used to be the case before the common wall was insulated. For this reason Mike would choose a detached house, although he does not want to lose the structure of his apartment (see Sect. 4.3). Another option, Daniel notes, is a corner apartment on the top floor to minimise the risk of noise from neighbours. Mike had the luck to find one.

The higher an apartment is located, the more beautiful the view. Daniel often feels locked up by his four concrete walls. An unobstructed view above the tree line, allowing him to see the city lights afar, could give some feeling of freedom.

In her fifth floor apartment, Pat very much enjoyed having an endless overview; she could keep on watching forever [M1b]. A garden somehow remains enclosed; it is finite wherever you look. For others, the height could come with disadvantages too. Maggie mentions her fear of fire, which relates to the need for exits [M4b].

Living on the ground floor in turn may cause problems of privacy [S2]. Mike would dislike it, unless his house is distant from the street. Norah has a front garden, but finds the intrusive looks of the people in the neighbouring care home sometimes disturbing. Elaine opted for a studio on the ground floor precisely because of its little garden. Also in other cases—apartments and houses—the garden is considered as very pleasant: a place to quietly sit outside, enjoy the sun [S1], welcome visitors, keep pets or dry laundry. When absent, however, it is not always missed. For Mike maintaining the garden would really pose a problem, forcing him to call in help. Ideal for him would be a small terrace with flower boxes and a seat: it requires little maintenance and yet allows him to sit outside. Tina pushes the idea a bit further to combine the best of both: most of all she would like an apartment with a beautiful view and a small roof garden.

4.3 Layout

The next theme introduced during the interviews concern the layout of the house: multiple smaller rooms or one big open space? John prefers open spaces, in part due to the fear of dark corners [M1b] he suffered from as a child. Pat's son Lucas does not dare to enter a room through a closed door either, because he cannot see what is behind [M1]. All doors are thus left open, except for the glass door to the living room. Rounding off corners might be useful to smooth the transition between spaces, John mentions. He also appreciates it if eating and sitting areas constitute one whole: when there are visitors he can keep himself in the background without having to leave the room [M4a]. According to him overview is important too [M1b]. This is why Maggie prefers not to have walls at all. Being unable to view the living room from the kitchen or vice versa is inconvenient. Moreover, the walls create very dark rooms where there are no windows, like the kitchen [S5]. Pat prefers one big space, but cannot stand always seeing everything, including the mess [S3]. Therefore she would install sliding doors she can close when needed. Where she would be able to live, for instance, is in a friend's loft which is especially furnished with built-in cupboards to store everything.

Mike, by contrast, clearly prefers separate rooms. He likes the structure of his current apartment: one living room with sitting area, eating area and desk area, plus a separate bathroom, kitchen and bedroom [M2]. He especially appreciates the fact that from his living room he can survey all other functional areas [M1b], but does not need to as he can simply close the door [S3]. He dislikes a big open house. He would either refurbish it or install partitions, also because of the size. As a child, Mike lived with his parents in a large town house. As the doors were always open, the entire 12 m long ground floor space and big garden behind were visible. This seemed way too big for him, especially in combination with the high ceilings. He prefers the enclosure of smaller lower rooms like the ones in his apartment [S7].

In the house of Elaine, Tina and Simon, different parts of the space are demarcated by furniture only. Simon could use a separate room to edit and store his photographs, but it is not a must. Daniel collected so many books that he had to turn his bedroom into a library where he sleeps on a mattress on the ground. In his ideal apartment, however, his library would be his living room, combined with an eating table and sitting area. To make it cosy, he would separate the library clearly from his workplace [M2]: a separate study would serve for administration and paperwork and accommodate his desk with four computers and all kinds of other stuff. In addition he would like one separate bedroom for himself, and one for when his nephews stay over.

The need for privacy [S2, M4a] also seems to influence how much space is needed and how it should be subdivided. Whereas Simon and Mike are happy with their house now, it would be very different when living together with someone. If Mike and his girlfriend lived together, concentrating on his work would become very difficult for him without an extra room. In fact, they would need two hobby rooms and two living rooms, which is unaffordable. When his son is visiting, Simon already starts having difficulties after 3 days. Before, when living together permanently, he missed that space to withdraw, so if he ever considers living together again, an extra room is definitely needed.

Also Pat needs that space to be able to 'run away from her family' [S2, M4a]. She and her husband chose the house because it has high ceilings and the first floor is spacious enough for a separate hobby and study room, with an extra bed to use in difficult periods. Smaller rooms than this would not work for her. And yet, a house can also be too big: David lives alone in his parents' house. Since all this unused surface must be maintained and heated, he is considering renting out part of his house. He prefers not to move because he is so used to living there. Elaine and Tina find it ideal that their apartment is not too big. Tina is not often at home anyway, and so cleaning takes less work. Maggie's children are at boarding school most of the time. Most of all she would like to live together with others, including her daughter, in a monastery, assuming that she still has her own room she can furnish [M4a].

4.4 Interior

In describing their house, participants mention hardly anything about the interior spontaneously. Apparently the furnishing is not always done purposefully. At Simon's and Mike's it is determined mainly by the outside world, also literally. All furniture in Mike's house has been donated by others, except for the desk chair he uses 99 % of the time. The TV came from his parents, the chairs from his girlfriend and the table too arrived at others' insistence. He used to eat at his desk and, before meeting his girlfriend, never had visitors, so buying a table would have made the interior more cluttered [S3]. Now he enjoys his meal more consciously, he notices, and he less often forgets to eat [M2]. Yet, without a relationship and his parents constantly wanting to adjust his house, he would not have nor need many of these things. His interior is a compromise between what he wants and what others want or consider normal. He has the feeling that he has to adapt to his interior instead of adapting the interior to himself. Since Simon has difficulties imagining what he wants, he relies on what he sees and tries to furnish his apartment the way he 'thinks it should be'. It is a hotchpotch of Ikea cupboards or shelves which he bought at a point when he needed them, mainly for practical reasons. Still, if he were moving, he would try to create the same interior again. After all, he has always lived in the same type of house and he is now used to it that way [M2].

Although the interior is precisely what one can decide about, not all participants live in the interior they like most. In a rental house, Tina explains, some things are impossible or nor allowed. Pat would like the interior of her previous house back: an austere cuboid with plain white walls and direct warm light [S3–5]. This was very 'quiet for [her] head'. Also, Simon would like his interior a bit more sober. He dislikes too many belongings, so an austere interior would please him more. Now he gets a feeling of chaos rather than structure. Maggie would like everything very open with pillows on carpets instead of seats and with big transparent boxes [M1a]. If Mike did not have to be considerate of others, he would put as little as possible in his interior because it distracts him. A friend of his lives in a small tent in his living room [S7, M4], with his mattress and sleeping bag, a small TV and internet connection. His clothes are put neatly into cardboard boxes with labels [S3]. Apart from that his apartment is empty. Mike would not go that far, but keep only his bed and desk, and invest in open bookshelves instead of a TV, chairs, a table and electronics.

Most important in Daniel's interior is his library, which he selected and implemented very consciously. In his previous house he combined grey metal shelves with colourful wallpaper. After moving to this apartment, he spent 8 years between grey walls and shelves. When, a few years ago, the shelves started to give way, he replaced them by beautiful blue Billy bookcases from Ikea, hiding the concrete walls behind the blue backs. Blue is his favourite colour, but the 2,000 books in his library make it into a very colourful whole. To keep it cosy, he uses different heights and little spotlights, each lighting a certain part of the library.

Some persons with autism like to order and display everything [M1b], John points out. In fact he displays most of his stuff too. Behind his bed a bookcase exhibits his CD's and DVD's. At a single glance he can get a complete picture and think 'it is a beautiful collection', without having to open his cupboard. His closed cabinet is very confusing for him, so he always puts clothes on the wrong pile. He wonders whether this would be equally difficult in an open wardrobe. Also Mike had problems keeping everything neatly sorted, and often forgot where everything was. Therefore he removed the doors from his cabinet and stuck labels every- where. Now he manages much better to keep everything ordered and keep an eye on it [M2–3]. In his kitchen, where he uses labels only, it is much messier. If he were able, he would remove all doors. For Pat's son, it would be better to have no doors on the cupboards or even no cupboards at all: as with other doors, he is afraid of what could be behind them. Pat would like, most of all, to hide everything in built-in cupboards in an austere white interior so that she would not have to see anything [S3–4]. Norah would like that too. David prefers closed cupboards because he does not need to see the mess, but they make finding things more difficult again [M1b].

5 Discussion and Conclusion

The stories of independent adults with autism reveal considerable differences in terms of living situation and housing type, ranging from living alone in a studio in the city centre to living with a family in a detached country house. In the search for a place to live, many choices need to be made for which many different arguments were brought up. Analysing these arguments sheds a more nuanced light on concepts of autism-friendly living environments. It shows that these concepts are not indisputable rules that can be applied straightforwardly, but can be interpreted in different ways and that one concept may reinforce but also counteract another. An open space may be preferred in order to avoid dark corners and increase predictability. In terms of controllability, an open space may provide overview and allow one to withdraw without having to leave the room. Another person may need separate rooms because they offer structure—a place for everything and everything taking place on its appropriate spot. The question then arises how far this subdivision of space should go in order to offer such structure. Moreover, other elements partly influence the preference for such a subdivision. A person may opt for multiple separate rooms because s/he just likes smaller enclosed spaces. Others cannot stand too many visual stimuli. In an open space you can see everything, including the mess. A space could be open if everything were stowed away in built-in cupboards as part of an austere interior. Yet this makes the environment less understandable for someone who prefers open cupboards with labels. These examples suggest that different concepts of autism-friendly architecture are not independent and that a balance must be sought in each particular situation. What is at stake, the stories suggest, is creating a living environment for *people* who have their own preferences, wishes, etc. Finding this balance will likely be easier when designing an environment for a single known inhabitant than when designing for multiple known or unknown potential inhabitants. In the former case, our study suggests, asking the right questions may already yield a lot of valuable information. Designers might even consider using the dilemmas as a tool to figure out the relative importance of certain concepts for the future inhabitant(s).

On the other hand, visits to participants' homes often gave a sobering impression: many of them are very ordinary houses where only details suggest that someone with autism is living there. Yet, if they were able to completely decide for themselves, some participants would live completely differently. The ideal situation they describe does not often correspond to reality. Because of their limited income, participants sometimes rent a social apartment. A certain fear seems to exist as to the quality of such apartments, in terms of acoustics, insulation, smoke, other inhabitants and even their robustness. Sometimes they are 'built sloppily', to use Daniel's words. Despite the limited possibilities, participants make the most of them to the extent possible. However, the ideal house remains, for most participants, utopian.

Their ideal situation starts not so much from 'how it should be (furnished)', but from how they would like it most. Living in a monastery may seem out-of-date, and yet it is Maggie's future dream. Most of all she would like to live in a monastery community with her own little place, where she earns her living by producing and selling things, and the interiors are open with pillows on carpets and transparent furniture. Mike perhaps would like to live like his friend: camping in his living room in a small tent with mattress, sleeping bag and as few belongings as possible. Daniel did furnish his house as he wanted, turning his bedroom into one big library. His family thinks he has to crawl over the books, but that does not bother him and he tries to make it as cosy as possible. Someone may thus feel perfectly O.K. in an environment that does not fit the traditional view of a good place to live. Our study thus contributes to a more nuanced understanding of concepts of autism-friendly architecture found in literature, but also to a more colourful image of what an autism-friendly living environment could be.

Acknowledgments This research was supported by the Research Foundation—Flanders (FWO), the European Research Council under the European Community's Seventh Framework Programme (FP7/2007-2013)/ERC grant agreement n° 201673, and the Department of Architecture. Thanks to the participants for sharing their time and experiences.

References

Ahrentzen S, Steele K (2009) Advancing full spectrum housing. Technical report, Arizona Board of Regents, Phoenix, AZ, US

Baumers S, Heylighen A (2010) Harnessing different dimensions of space. In: Langdon PM, Clarkson PJ, Robinson P (eds) Designing inclusive interactions. Springer-Verlag, London

Bogdashina L (2003) Sensory perceptual issues in autism and asperger syndrome: different sensory experiences—different perceptual worlds. Jessica Kingsley Publishers, London

Bollnow OF (2011) Human space. Hyphen Press, London

Brand A (2011) Living in the community. Housing designs for adults with autism. Available at: www.hhc.rca.ac.uk/CMS/files/Living_in_the_Community.pdf. Accessed 4 Nov 2013

Coolen H (2006) The meaning of dwellings. Hous Theory Soc 23(4):185–201

Grandin T (1995) Thinking in pictures and other reports from my life with autism. Doubleday, New York

Mostafa M (2007) An architecture for autism. Archnet-IJAR 2(1):189–211

Mostafa M (2010) Housing adaption for adults with autistic spectrum disorder. Open House Int 35(1):37–48

Sanchez P, Vazquez F, Serrano L (2011) Autism and the built environment. In: Williams T (ed) Autism spectrum disorders—from genes to environment. InTech, Croatia

Vogel C (2008) Classroom design for living and learning with autism. Asperger 3:30

Part V
Collaborative and Participatory Design

From Designing for the Patient
to Designing for a Person

M. Annemans, E. Karanastasi and A. Heylighen

Abstract Research on inclusive design stresses the value of user experience as a resource to design with respect for the diversity in human abilities and conditions. So far, however, relatively little research has been conducted on how exactly user experience benefits design processes and their outcome. How is it introduced into the design process, what kind of knowledge do designers get from it and how does it inform and direct their design process? The study reported here addresses these questions in the context of a design studio in which student architects designed a Maggies Cancer Caring Centre. After briefly discussing the role of (user) experience in design processes, we sketch the context of the Maggies Centres and introduce the assignment and procedure of the design studio. In order to analyse how different sources of information about user experience feature in students design process and outcome, we rely on documents students handed in, notes taken and audio recordings made during conversations with patients and care givers and students presentations. Four sources of information about user experience were addressed explicitly or implicitly by various students: direct communication with cancer patients and with people working at a day care centre; the person of Maggie Keswick; the architectural brief and exemplary projects of user-sensitive buildings. Despite its limitations, participation in this studio clearly increased students' knowledge on specific users. Many students mentioned the fact that a Maggie's

M. Annemans (✉)
Department of Architecture, Research[x]Design, University of Leuven (KU Leuven),
Kasteelpark Arenberg 1/2431, Leuven, Belgium
e-mail: margo.annemans@asro.kuleuven.be

M. Annemans
Osar Architects nv, Antwerp, Belgium

E. Karanastasi
Department of Architecture, University of Leuven (KU Leuven), Leuven, Belgium

A. Heylighen
Department of Architecture, Research[x]Design, University of Leuven (KU Leuven),
Leuven, Belgium
e-mail: ann.heylighen@asro.kuleuven.be

P. M. Langdon et al. (eds.), *Inclusive Designing*, DOI: 10.1007/978-3-319-05095-9_17,   189
© Springer International Publishing Switzerland 2014

Centre should not be designed for the patient but for a person. If only this insight remains, it will already contribute to them becoming architects who design with more than just functionality in mind. Additionally, the existing Maggie's Centres provided students with examples of exceptional architecture. The studio assignment thus drew their attention to the possibility to create extraordinary buildings, appealing to users and specialists alike, designed for the well-being of everyone involved with them. By doing so it opened students' eyes to designers ability to really transform the daily lives of the people engaging with the spaces they conceive.

1 Introduction

Research on inclusive design stresses the value of user experience as a resource to design products and environments that respect the diversity of human abilities and conditions. Elaine Ostroff (1997) therefore introduced the term user/expert, denoting '*anyone who has developed natural experience in dealing with the challenges of our built environment*'. Since user experience may offer designers unique insights (e.g. Pullin 2009), several methods for involving it in design are developed, extending traditional focus group interviews into more embodied approaches (e.g. Annemans et al. 2012a; Heylighen 2012), critical user forums (Dong et al. 2005; Cassim 2007) and co-design (e.g. Tsianakas et al. 2012). So far, however, relatively little research has been conducted on how exactly the user experience brought in through these methods benefits the design process and its outcome. How is it introduced, what kind of knowledge do designers get from it, and how does it inform and direct their design process?

This paper addresses these questions in the context of a design studio where student architects designed a 'Maggie's Cancer Caring Centre'. These centres, of which 14 are operational so far, are meant to improve the wellbeing of people affected by cancer. Based on the belief that high-quality architecture may support people's wellbeing, the Maggie Keswick Jencks Cancer Caring Centre's Trust puts users' (spatial) experience at the centre of the design process. For cancer patients, stress and anxiety are frequent but have highly context and person specific causes; designing for them thus requires that designers consider their particular concerns, wishes and experiences (Mullaney et al. 2012). In designing a Maggie's Centre, world famous architects like Zaha Hadid or Richard Rogers were challenged to work with and for specific users. By studying how various information sources on user experience impact student architects' design of a Maggie's Centre, we aim to gain insight into what knowledge of people and users (student) architects use in their design and how it informs and directs their design process.

After briefly discussing the role of (user) experience in design processes, we sketch the context of the Maggie's Centres and introduce the design studio's assignment and procedure, and analyse how different sources about user experience feature in students' design process and outcome. Finally, we confront our findings with literature on experience in design processes, and formulate lessons learned to deepen (student) designers' understanding of real persons engaging with their design.

2 (User) Experience in Design

In traditional societies, where human-made objects were conceived, made, and used by the same person (Jones 1970), the experience of using the object could be fed back directly in the design and making of its material, physical features. The industrial revolution introduced a separation between the designer (who conceives an object), maker (who produces it) and user (who experiences it). As a result, the direct feedback loop got interrupted. Today, designers typically conceive products and environments with an eye to offering users a certain experience, without having direct access to their motivation, values and prior experiences. How users eventually experience the result may correspond to what the designers intended but might also differ from it in various ways (Crilly et al. 2008). Inclusive design's emphasis on involving user experience in the design process can be understood as an attempt to bridge this gap.

Research on inclusive design advocates involving user experience in the design process, in line with user-centred design (Dong et al. 2003). Adopting a design approach in which the actual people being designed for and their real-life experiences are present, is considered crucial if the resulting design is to benefit people of different ages and abilities. The idea is to involve real people who actually take part in designing, contributing to the design process from their own personal experience (Dong et al. 2005; Cassim 2007; Pickles et al. 2008; Mullaney et al. 2012), giving input and reflecting on solutions proposed by the designers (Tsianakas et al. 2012) or even proposing ideas themselves (Luck 2012).

In practice, however, involving users during design is considered time consuming and thus expensive (Dong et al. 2003). Designers therefore rely mainly on other forms of experience, offering ersatz feedback on how future users will experience the product or space being designed. Architects, for instance, rely heavily on their personal experiences of places they have visited (Downing 2000), on exemplary buildings designed by others in books or magazines and on projects they have designed themselves (Heylighen and Neuckermans 2002). Throughout their career they collect an extensive record of precedents, serving as a source of knowledge during design. Moreover, through engaging in various social situations and interactions, (student) designers acquire a 'culture medium', which embraces various substances, phenomena and traces, from both within and outside design, all of which can function as raw material for design (Strickfaden et al. 2006). As will become clear in the next section, designing a Maggie's Centre potentially combines these different forms of 'experience' in architects' design processes.

3 Maggie's World

Maggie Keswick was a landscape designer. The importance of a supportive environment for her emotional wellbeing became particularly clear to her when she was told that the cancer she had been battling before had returned and she had only a few months left to live. She remembered the announcement as follows: '*How long have we got? The average is three to four months ('and I'm so sorry, dear, but could we move you to the corridor? We have so many patients waiting...')*' (Keswick and Jencks 1995). The corridor she was moved to can be imagined by everyone who ever visited a hospital. Corridors, toilets and waiting areas are the main hospital spaces for which Maggie advocated the provision of alternatives: '*waiting areas could finish you off*', they do not support you as a patient but rather tell you: '*How you feel is unimportant. You are not of value. Fit in with us, not us with you*'. She was convinced that with little effort the opposite could be achieved (Keswick and Jencks 1995).

Based on Maggie's experiences and initiated by her and her husband Charles Jencks, the Maggie's Centres aim at creating supportive environments that add to their users' wellbeing. Starting from *A view from the frontline* (Keswick and Jencks 1995), a booklet about Maggie's personality and how the disease affected her entire being, the Trust governing the centres wrote an architectural brief for their design. Unlike most briefs, it focuses on the creation of spaces for different moods and uses rather than on square metres or number of rooms (Trust 2011). Architects are expected not so much to translate rules into spaces, but rather to think along and come up with a truly inspirational building that suits the needs of patients, relatives and personnel: '*So we want the architects to think about the person who walks in the door. We also want the buildings to be interesting enough that they are a good reason to come in rather than just 'I'm not coping*' (Trust 2011).

For certain spaces, the brief lists more specific requirements. A Maggie's Centre should be approximately 280 m^2, the only numerical value in the brief, and contain an entrance, sufficient office space, a kitchen and lavatories. For each space the atmosphere aspired to is described without prescribing a fixed solution. The entrance should be welcoming, not intimidating. Unlike what is often the case in a hospital, the layout should be clear and the building as light as possible. The lavatories should not be all in a row with gaps under the doors, but private enough to cry in. Apart from descriptions of specific spaces, there are also pointers regarding the overall architecture. The Maggie's Centres and the way they are designed should raise your spirits, be safe and welcoming but not too cosy, and increase people's sense of connectedness (Trust 2011).

4 The Design Studio

Maggie's story combined with the specific brief and examples of existing centres, inspired us to set up a design studio for student architects. The 34 master students attending the studio (15 female, 19 male) were asked to design a Maggie's Centre for Leuven. The studio was led by two professional architects (including the second author). Students received the brief formulated by the Trust and a plan of an area near the university hospital, where they could choose their own spot to situate their project. The area has an advantageous slope and alternating areas of dense thicket and deforested spots.

Students also received various other sources: the first author guest lectured about how users experience Maggie's London, pointing out multiple levels of emotional impact of the built environment (Annemans et al. 2012b); other guest lectures addressed the subjectivity of spatial experience, or post-traumatic stress in patients diagnosed with cancer and intervention techniques related to space; students participated in a workshop with three (ex-)cancer patients testifying about the importance and character of healing environments, based on their subjective experience; they visited a daycentre for patients with life threatening diseases; and they analysed in groups an existing Maggie's Centre.

Finally, every student presented his/her project for a jury of two studio teachers and two guest lecturers (including the first author). Seven projects were presented to two of the three (ex-)patients and an oncologist working in the university hospital. This was expected to sensitise students to differences between architects and lay persons in reacting to or dealing with the presentation of design ideas.

5 'Maggie' in the Design Process

The design studio aimed to raise students' awareness of the diversity in people's (*c.q.* 'cancer patients') experiences and sensitivities. Yet, how present were these people in the design (process)? And how did students refer to the people using a Maggie's Centre? We analysed documents students handed in (drawings, 'storyline panels', inspiration sources), notes taken by the authors during the presentations and audio recordings of the final presentations and conversation with patients and care givers. We also looked at how these people were (re)present(ed) during the design process. Four sources about user experience were addressed explicitly or implicitly by various students when (re-)presenting their design. The first and most straightforward information came from the direct communication with cancer patients, and people working in the daycentre. Testimonies by patients triggered students' awareness of the specificity of the group they were designing for, but also of the diversity within this group which they otherwise might have considered as 'patients'. Second, the person of Maggie Keswick was very much present during students' design process. Although she spoke to them through a written source only, her message came through quite strongly. Maggie took the role of representing all unknown users, still being a real person, in a real situation, with strong ideas on her medical treatment, space, personal empowerment and even nutrition. Third, there is the architectural brief, underlying the assignment, but also translating user needs into a more architectural language. Finally, as world famous architecture forms an inspiration source for many (student) architects, user-sensitive examples of other Maggie's Centres or examples of architects designing sensory-rich spaces, seemed to add to the user-related qualities of students' designs.

5.1 Interaction with Real-life People

Many students explicitly mentioned the dialogue with (ex-)cancer patients as an important source of inspiration and information at different stages in the design process. During the site visit, they explored the given terrain with this dialogue in mind. One student chose his centre's location away from the hospital, at the most quiet place, based on what the patients had said: '*From the talk with the user/ expert, I derived that they expect from a Maggie's Centre that it creates a whole new living atmosphere, not closed off, but visually separated from the hospital. Therefore, I chose this spot in the woods, away from the hospital, with a buffer formed by the relief and the vegetation, accessible from the other street*'.

Also while designing, the patients' personalities were never far away. A student cited in his presentation a specific quote from a patient. She had said: '*During my treatment, it was very hard for me to concentrate. Reading a book was not possible*'. Obviously reading books is not a patient thing, but an aspect of this woman's personal life. In his design this student provided a quiet room, not just for isolation but specifically designed to be able to listen to music, or as he explicated '*an audio book, since it is hard for them to read*' (Fig. 1).

Not all students directly linked their design decisions to a specific element or quote. Some spoke in more general terms about the users' influence, like: '*The workshop with user/experts made us feel the difficulty and the nuance which we would have to use in the assignment*'. While it is hard to pinpoint exactly which design aspects stemmed from this understanding, the project testified to the students' sensitivity about the patients' wellbeing. Interpreting the client's wishes and desires is a task of an architect; here too, someone translated the patients' need to be able to retreat into the central concept of her design. A structuring object such as an equipped wall became a meaningful element to enable users to '*disappear into the closet*' when needing time for themselves.

5.2 Maggie

Like any other architect asked to design a Maggie's Centre, students were provided with the booklet *A view from the frontline* (Keswick and Jencks 1995), in which Maggie tells her story of being diagnosed with cancer and how she, a landscape designer and mother interested in Eastern medicine and meditation, experienced her environment throughout this process. It provides user information in a passive, one directional way, but many students found it inspiring. As mentioned Maggie

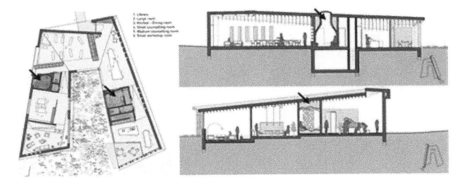

Fig. 1 Quiet rooms, specifically designed to listen to e.g. an audio book (© Pieter-Jan Debuyst)

addressed waiting rooms, hallways and toilets as most depressing spaces in the hospital. During the presentations, a student literally told how, at the beginning of her design process, she worked in a very functionalist way: only when she went back to Maggie's story (and the movies on the Trust's website), did she realise she should take a different approach, so she redesigned her centre into a building without hallways, where dynamic spaces followed one another. Even more explicit was someone who showed a painting by Claude Monet of a woman walking in a field, saying that she was how he imagined Maggie. While presenting for the patients and oncologist, the same student called the people who would use his building '*his Maggies*'.

For some Maggie's personal spirit of enjoying life to the fullest was even a starting point. One student showed, as first slide of her 'storyline panel' an image with the saying '*Today is a good day*'. Also the message that the building should be anything but a hospital came across. Some based their design on the archetypical terrace house, others came up with a resort-like typology. Either way, not having waiting areas and hallways was a central theme for many. The retreat of the toilet was given alternatives or was upgraded with daylight and some more space to move.

5.3 The Architectural Brief

The brief of the Maggie's Centres is somewhat different from the usual case, focussing more on atmosphere than on square metres. Still it remains the closest to what (student) architects are used to starting their design from. Most of the designs feature elements mentioned in the brief, e.g. flexibly usable spaces, a central kitchen island and table, therapy and lecture rooms, spaces to retreat and collective areas. Given Maggie's fascination with nature, the presence of green was an essential element in the centre's quality. Also the amount of natural elements on the given location became an important constituent of many projects. Some chose

Fig. 2 Natural (*top left*) and designed (*top right*) green environment (© Laura Van Bel; Pieter-Jan Debuyst). Design based on the archetype of the house (*bottom*) (© Matthias Salaets)

to make their building disappear in the wood, or reflect it so as to become as transparent as possible; others really worked with it, designing the green just as they designed the building (Fig. 2).

Despite requiring these clearly listed elements to be present in each centre, the brief also challenges architects to not follow it blindly, but make spaces that help the transition from being patients, or even cancer victims, to becoming individuals again. It even dares designers to come up with maybe contradictory things (The Maggie Keswick Jencks Cancer Caring Centres Trust 2011). Whereas none of the students did refer to this requirement explicitly, several seem to have taken up the challenge. By designing a longitudinal building, some questioned the notion of centrality of the kitchen, for example. How do you make a central kitchen and avoid hallways when all spaces are located in a row? This may not be easy but choosing this spatial configuration brings all the rooms closer to nature, thus contributing to more users' wellbeing. One student also explored the meaning of '*domestic space*'. Do users experience the archetype of a house as domestic, despite a rather abstract material choice?

5.4 Maggie's Centres and Other Built Examples

Consciously or not, both professional architects and students build on their knowledge of exemplary architectural projects (Heylighen and Neuckermans 2002). Given the list of famous international architects who preceded the students in designing a Maggie's Centre, it is likely that they derived a source of inspiration from them. Images from other centres were explicitly displayed on students' panels. Amongst others, the differentiated light levels in OMA's centre for Gartnavel formed a popular reference. By referring to this example, students almost automatically addressed two requirements formulated in the brief, namely the presence of light and providing intimacy when needed. However, students did not limit themselves to Maggie's Centres to find inspiration on user-sensitive architecture. The work of Swiss architect Peter Zumthor was frequently cited as an example of architecture relating to nature without neglecting the atmosphere inside. Only one student specifically looked for examples of care buildings focussing on user experience. He stated that the Ronald McDonald family room, a facility for families of hospitalised children in the Netherlands, showed him how to deal with wellbeing in relation to the built environment.

While the actual design outcome may differ considerably, there are only so many typological ways of dealing with inside-outside relationships combined with the required programme. No wonder some student projects reflected existing Maggie's Cancer Caring Centres, maybe even unintentionally. This could be observed even on a more general level. Typologies such as a beam like building or archetypical houses in a uniform material can also be considered basic architectural references.

6 Discussion

If we confront the sources about user experience documented above with the interpretations of 'experience' in the design process described in literature, some additional sources can be identified. Indeed, (student) architects rely on their own bodily experience of places they have visited when designing. The central kitchen table is referred to by different students as a table at the popular bread and breakfast location Le Pain Quotidien, not designed by world famous architects, but definitely creating common ground with the (ex-)cancer patients attending the final presentation. When asked what they liked most, this was the example patients cited. The importance of buildings or spaces designed by others, which students know from literature or courses, or are advised to look at by studio teachers, is illustrated above. We could not clearly identify references to other buildings students designed themselves. Yet, one teacher referred to an assignment the year before whereby students designed an apartment building. That some students extruded a plan to a height of 3 m instead of designing in 3D, he ascribed to this assignment, which would have raised their interest in piling up identical floor plans.

In reality, the different sources of user experience involved in design (processes) are not as clearly distinguishable as presented here. Besides the real-life people invited for the students to talk with and the person of Maggie, a variety of in-betweens may have informed the design as well. The Trust's website contains movies showing testimonies by the centres' users. Here too, people somehow affected by cancer, give a personal, often touching, view of how the organisation and building add to their personal wellbeing. Before the studio started, students received a list of movies they could watch to become familiar with the life world of the people they would design for. The design teachers mentioned both testimonies and movies as important references for students. One student even incorporated testimonies from the website in his final presentation, making it seem as if the people visiting the centre commented on his design.

While introducing users in a design process is not new, confronting students with real people is not a common practice in our programme. Instead teachers or students 'invent' their buildings' users, adapting them to their design instead of vice versa. In this design studio people representative of these users were involved, but they did not wield much power, as is often the case (Cuff 1989). The (ex-)patients and oncologist who participated were invited at the start of the studio and at the final presentations of some projects, selected by the teachers. Yet giving feedback along the way and grading the projects was done by the teachers, trained in architecture and design.

For the coming year, we have slightly altered the approach. Small groups of 2–3 students are motivated to engage with 'their' user/expert and discuss their design with them along the way, while, sharing their findings with the other groups, so as to get access to a more diverse set of perspectives on cancer care in the broadest sense.

7 Conclusion

If designers are to design for inclusion, informing them about peoples' experiences is a key concern. By analysing the use of different sources on user experience in a design studio, we gained a better understanding of what kind of knowledge, students refer to in their design and how it informs and directs their design process. For many students these sources functioned as something to fall back on when they were stuck while designing. The presence of real persons, representing possible users of the centre, at both beginning and end of the design process, challenged students not to forget about them, not while designing, not when presenting. Still, with the real users being an audience rather than a source of feedback, nuances, like the shades between patient and person were not always taken into account. With the altered approach of this year's studio, we hope to improve students' sensibility towards the future users of the buildings they design.

Since in the studio reported on the people representing 'the users' were not present in person during the entire period, the different sources about their experience should be compatible, at least to some extent. In spite of small nuances, we indeed found many similarities in the topics addressed by the patients, the oncologist, Maggie's booklet, the brief and the built examples. For example, the presence of nature, pointed out by Maggie as a crucial element, and thus included in the brief, was also appreciated by the patients and oncologist during the final presentations. Whether this nature should be 'wild' or 'designed', depends on personal opinions. The same is true for the small isolation spaces. Although the oncologist thought they would hardly be used, the patients could imagine retreating in there, alone or with a companion. This kind of small inconsistencies, or nuanced interpretations of elements mentioned in the brief, challenged students to question the assignment and the actual meaning of wellbeing for different persons.

Despite the limitations discussed above, the assignment offered a unique opportunity to study how users and user experience can enter the design process through different means. As studying and passing on experience is not easy, it is important to start growing awareness of the subject during education, especially when aiming to design for wellbeing. Participation in this design studio clearly increased students' knowledge about specific users. Many students mentioned the fact that a Maggie's Centre should not be designed for 'the patient' but for a person. If only this insight remains for their future careers, it will already add to them becoming architects who design with more than just functionality in mind. Additionally the existing Maggie's Centres offered students examples of exceptional architecture. The assignment thus drew their attention to the possibility of creating extraordinary buildings, appealing to users and specialists alike, designed for the wellbeing of everyone involved with them. By doing so, it opened students' eyes to designers' ability to really transform the daily lives of the people engaging with the spaces they conceive.

Acknowledgements This study received support from the Institute for the Promotion of Innovation through Science and Technology in Flanders (IWT-Vlaanderen) through a PhD grant of the Baekeland programme, osar architects nv, and the European Research Council under the EU's Seventh Framework Programme (FP7/2007-2013)/ERC grant agreement n° 201673. Thanks go to all involved in the studio, including Mauro Poponcini, Hans Verplancke, the guest lecturers, TOPAZ, Stichting tegen Kanker, the user/experts and the oncologist.

References

Annemans M, Van Audenhove C, Vermolen H, Heylighen A (2012a) Hospital reality from a lying perspective. In: Langdon PM, Clarkson PJ, Robinson P, Lazar J, Heylighen A (eds) Designing inclusive systems. Springer, London

Annemans M, Van Audenhove C, Vermolen H, Heylighen A (2012b) What makes an environment healing? In: Proceedings of 8th international design and emotion conference, London, UK

Cassim J (2007) It's not what you do, it's the way that you do it. In: Stephanidis C (ed) Universal access in HCI, Part I, HCII 2007, LNCS 4554

Crilly N, Maier A, Clarkson PJ (2008) Representing artefacts as media: Modelling the relationship between designer intent and consumer experience. Int Journal Des 2(3):15–27

Cuff D (1989) The social art of design at the office and the academy. JAPR 6(3):186–203

Dong H, Keates S, Clarkson PJ, Cassim J (2003) Implementing inclusive design. In: Carbonell N, Stephanidis C (eds) User interfaces for all, LNCS 2615

Dong H, Clarkson PJ, Cassim J, Keates S (2005) Critical user forums. Des J 8(2):49–59

Downing F (2000) Remembrance and the design of place. A&M University Press, Texas

Heylighen A (2012) Inclusive built heritage as a matter of concern. In: Langdon PM, Clarkson PJ, Robinson P, Lazar J, Heylighen A (eds) Designing inclusive systems. Springer-Verlag, London

Heylighen A, Neuckermans H (2002) Are architects natural case-based designers? Des J 5(2):8–22

Jones JC (1970) Design methods: Seeds of human futures. John Wiley, Chichester

Keswick M, Jencks C (1995) A view from the frontline. Maggie Cancer Caring Centres

Luck R (2012) Kinds of seeing and spatial reasoning. Des Stud 33(6):557–588

Mullaney T, Petterson H, Nyholm T, Stolterman E (2012) Thinking beyond the Cure. Int J Des 6(3):27–39

Ostroff E (1997) Mining our natural resources: the user as expert. Innovation, 16(1):33–35

Pickles J et al (2008) Experience based design. Clin Gov 13(1):51–58

Pullin G (2009) Design meets disability. The MIT Press, Cambridge

Strickfaden M, Heylighen A, Rodgers P, Neuckermans H (2006) Untangling the culture medium of student designers. CoDesign 2(2):97–107

Trust (2011) Maggie's architectural brief. The Maggie Keswick Jencks Cancer Caring Centres Trust

Tsianakas V, Robert G, Maben J, Richardson A, Dale C et al (2012) Implementing patient-centred cancer care. Support Care Cancer 20(11):2639–2647

Accessible Participatory Design: Engaging and Including Visually Impaired Participants

C. Andrews

Abstract This paper provides an overview of various participatory design techniques made accessible for visually impaired participants. Within a project to design a tactile, pedestrian, navigation aid various participatory techniques were utilised. A suggestion for application is made alongside points of interest for each technique which may be of value to researchers new to both participatory design and design for visually impaired users. The conclusive paragraph draws out generalizable findings alongside the notion that, as the designers empathy for the user group increases, so to will their ability too create accessible participatory approaches.

1 Introduction

Whilst laying out participatory design (PD) as a separate technique from user-centered design (UCD), Sanders (2002) defines UCD as 'designing *for* the users' whilst PD is defined as 'designing *with* the users'. This adds another level of difficulty to the design process as the challenge of how to make it accessible and enjoyable for the non-design-based participants must also be considered.

Brandt (2006) notes that 'designing the design process itself is just as important as designing the artefact', and in the context of accessible products and inclusive design this is even more prevalent. As the design process is often visually driven: from early stage mind maps and mood boards to sketching, prototypes and information layouts, 'engaging and involving' (Sanders et al. 2010) visually impaired (VI) participants will involve considerable methodology planning and deliberation.

C. Andrews (✉)
The National Centre for Product Design and Development Research (PDR),
Cardiff Metropolitan University, Cardiff, UK
e-mail: clandrews@cardiffmet.ac.uk

P. M. Langdon et al. (eds.), *Inclusive Designing*, DOI: 10.1007/978-3-319-05095-9_18,   201
© Springer International Publishing Switzerland 2014

Sanders et al. (2010) overviews various documented PD techniques and categorises them as:

- 2-D collages
- 2-D mappings
- 3-D mock-ups
- stories and story boarding
- diaries
- cards to organise, categorise and prioritise ideas
- game boards
- props
- participatory envisioning and enactment
- improvisation.

These techniques are playful in approach which encourages engagement in the process. They also have a non-direct element about them; the techniques focus on understanding tacit knowledge rather than the explicit knowledge gained from traditional methodologies (Sanders 2002). They differ from traditional UCD research methods such as focus groups, interviews and observation as the participant is much more integrated into the task. Rather than a question–response relationship, the participatory methods ask the participants to become creative in their own right.

There are some excellent open resources giving new researchers tips on how to effectively conduct focus groups and interviews with VI participants (Gerbe 2000; Henry 2003; Kroll et al. 2007). These sources give valuable information on a variety of issues a researcher may encounter when working with VI participants including: suggestions on room layout, communication, consent, location and preparation. However, there is little collated information available on how to approach more interactive participatory methods with VI participants.

This paper intends to overview some participatory techniques accessible to VI participants that have been utilised by the author in the hope that future researchers may find the information of value when planning their own studies.

2 The Study

The majority of the techniques described were employed throughout the author's PhD studies. The brief set was the research and development of a pedestrian handheld navigation device with tactile output. The device was not intended to be specifically a mobility aid for VI users but to be a mainstream navigation aid. VI people were asked to participate in the design process for their expert knowledge of the tactile sense and to ensure the final product would be accessible to VI users.

2.1 Recruitment

Participants were originally sourced through three organisations dedicated to aiding VI people: Cardiff Institute for the Blind (part of RNIB), Royal National College for the Blind and Action for Blind People: Technology Department (part of the RNIB). The participants who engaged most with the project were those sourced through Action for Blind People which highlighted the value of finding participants who have an interest in the product being produced. Their prior involvement with the Technology Department marked them out as actively interested in technology and genuine potential users; consequently, they invested more into the design of the product and so engaged more in the participatory process. It also meant that their knowledge of existing products and available technologies was far superior to other involved participants. The Action for Blind People focus group consisted of five members, three male and two female, aged from 27 to 35; the groups were video and audio recorded, the audio was transcribed and coded to uncover recurring themes and opinions and the video utilised to analyse physical movement (such as product exploration techniques).

3 Collages

As stated by Sanders (2010), 2-D collages can be utilised throughout a PD process to probe for existing knowledge, prime the participants in the area of interest, understand viewpoints and emotion or generate ideas. For the project in hand, 2-D collages were created in the form of moodboards to understand the aesthetic and ergonomic factors of existing products. These allowed the designer to have visual prompts when sketching to aid development of the design. It was essential that these moodboards be created by the participants rather than knowledge assumed by the researcher; however, without the ability to create physical moodboards the process was edited to be verbally accessible.

3.1 Collage Method

A group of four participants was gathered, the concept of moodboards was explained alongside the aim of the moodboard in hand, in this case 'to document physical attributes of small electronic devices that either aid or hinder access for VI users'. Participants were then asked to 'call out' product features and how they help or hinder; an example response is 'Virgin Media TV remote buttons, bad because they're too shallow'. Responses are recorded and a physical representation is produced post-event to allow for easy presentation of information to designers.

3.2 Points of Interest

3.2.1 Engagement

The value of 2-D collages largely stems from the creativity element which engages the participants; conducted verbally, these methods could easily take place as a straight forward focus group discussion. However, asking the participants to call out their answers adds an element of chaos not regularly found in formal focus groups, which was deemed to be fun and engaging for the participants.

3.2.2 Dominating Participants

As with any group work scenario, full participation may be hindered by dominant individuals. It is the researcher's role to ensure this does not affect the output of the task. This is of particular relevance with VI participants as they cannot utilise the visual cues of somebody who may be waiting to say something. The challenge lies in controlling this whilst still allowing the chaotic element (as mentioned in Sect. 3.2.1) which helps engage participants. A suggestion to rectify this is that the researcher calls participants names at random and they must give an answer on the spot: this allows for the fast thinking, game-like approach to be continued but in a more controlled manner.

3.2.3 Reliance on Prior Knowledge

Whilst creating moodboards with sighted participants, no prior knowledge of the products need be available; the participant can simply be handed a magazine and asked to cut out inspiring pictures. As VI people must rely on their prior knowledge, to gather a broad range of answers the researcher must encourage broad thinking and out of context examples that may still be applicable.

3.2.4 Focus on Detail

As VI people explore detail before overall shape it might be small detailing that draws them to the object, rather than the overall shape or look. It may be simply 'the layout of the buttons' or the clarity of orientation and this must be represented properly in the final visual boards. The reverse is also true: that they enjoy the form of the product but are not able to access the function.

3.2.5 Visual Terminology

VI participants, regardless of their visual ability, regularly use visual terms, such as 'it looks beautiful' to describe objects. Care must be taken to fully understand what is meant when a VI participant describes something in a visual manner as fully sighted individuals may understand the statement differently.

4 Foam Models

Models and prototypes enable participants to discuss form and feature placement in a manner that may be difficult to do verbally. The challenge within PD is to make this accessible and fun to the participants. Keeping models low fidelity and 'sketch like' removes pressure to perfect the models and allows the participant to enjoy the task regardless of skill.

4.1 Foam Modelling Method

Many blank shapes were created in Styrofoam to express a selection of potential overall forms for the product (based upon previous discussion and tasks). The participants were then asked to explore the models and dictate to the researcher where they would envisage key features of the product; these features could then be drawn on creating a very low-fidelity model of the product. The participants were also able to feedback on the overall shape and grip and choose preferred forms which would aid the designer in concept selection and development.

4.2 Points of Interest

4.2.1 Opposite to Usual Product Exploration

As stated by Miao et al. (2009), VI people will first explore detail then build up a mental model of the product. In this task, the researcher is asking them to do the opposite and feel the overall shape before they develop detail. The participants seemingly enjoyed and engaged with this task as it allowed them to add something tangible to the design.

4.2.2 Fully Explain the Prototype Stage

If the prototype is at the initial stages of design this must be made clear to the participants to stop any frustration at lack of detail. Though this would also be true with sighted participants, it is heightened with VI participants as they may not be able to perceive the many visual cues that might give away the stage of development. At later stages in the process this becomes even more prevalent: as modelling materials may be both visually and tactilely very realistic, the participants must be verbally assured that they are allowed to be honest and changes are still possible based upon their feedback.

4.2.3 Model Strength

The models will be explored through touch and so should withstand vigorous tactile exploration. If the model is delicate, a VI participant will often have no warning of this until it is too late as the force of the tactile exploration itself may be enough to break it.

4.2.4 Material Selection

Material selection should be appropriate for the level of development the product has seen. Foam worked well as a material for feedback as whilst it kept the overall shape, which allowed participants to experiment with different grips and orientation easily, it is also editable by the participants on a surface through applying pressure with their fingers or with a pen.

4.2.5 Do Not Be Precious

Though many hours may have been spent shaping and moulding the models, participants with limited experience of model making (which VI people are likely to be) may not recognise this; the goal of the exercise is to allow the participants to both feedback on and edit the design and so they should be encouraged to do so as much as necessary to communicate their views.

4.2.6 Avoid Tactile Noise

Any tactile elements that are not in direct focus or consideration for the participant should be highlighted at the beginning of the task. As VI participants utilise minor details to build up their mental models, incorrect minor details even on low-fidelity models can be of high significance. This concept is also documented by Miao et al. (2009) when creating paper prototypes of computer interfaces for VI user trials.

5 Cards

Separating features, ideas, themes out onto cards allows participants to 'organise, categorise and prioritise' concepts (Sanders 2010). Cards are very easily presented in a verbal manner to a VI audience with a variety of applications.

5.1 Card Method

A simple successful method to utilise cards is to aid resolution for decisions that cause debate within focus groups. Features, functions and qualities are separated out to create 'cards'. The researcher presents two or more cards verbally to the participants and asks them to immediately and instinctively respond to which is more important. Examples include: 'Features or Price?' or 'Tactile output, visual output or audio output?'

5.2 Points of Interest

5.2.1 Visual Aids for the Researcher

Whilst the cards will be presented verbally for the VI participants, visual prompts aid the researcher in quickly presenting results and knowing what question will come next. For the project in question, the author simply used a word processor and read from the screen as a prompt for the verbal presentation and copied/cut and pasted the cards to display results; in hindsight, physical cards (of which photographs can be taken for examination post-task) might have been quicker and less obtrusive as the clattering of a keyboard and mouse can be distracting.

5.2.2 Present Limited Cards at Any One Time

With no visual cues, organisation of multiple cards is more difficult as it relies on memory or regular prompting by the researcher, which can be tedious for both parties. Present only limited cards for more clear and accurate responses from the participants. Plan the order in which the cards will be presented beforehand to ensure conclusions are reached in the most effective manner.

5.2.3 Do Not Rely on Memory

As previously stated, having the researcher verbally list the cards for each question becomes tedious for both the researcher and the participant. There is no way to tell without extended discussion whether the participant has remembered all the cards and so results will be less valid.

6 Existing Product Feedback

Existing products with similar features to the product being designed were utilised to gain feedback on aesthetic and ergonomic design. These products are not necessarily direct competitors, merely products with similar features to the product being designed.

6.1 Product Feedback Method

Relevant existing products were placed on the table in front of the participants who were asked to pick each one up and explore it. Participants were asked what features of the product they liked or disliked.

6.2 Points of Interest

6.2.1 Avoid the Props' Existing Use

It may take some time for VI participants to recognise the existing product or in some cases they may not recognise the product at all. Whilst it is interesting to see how someone who has no prior contextual knowledge interacts with an object, avoid the notion of 'testing' the participant as it can be frustrating and patronising. In the author's experience, the participants would first explore the product before asking the function, which allowed time for the initial exploratory acts to take place without any context.

6.2.2 True Feedback on Form

VI participants are in a good situation to give true feedback on ergonomics and form as they are not in a position to be influenced by graphical cues indicating orientation or use. To make the most of this, the researcher must take care not to influence the participant unintentionally; examples of this include: the researcher passing the object to the participants indicating orientation and the researcher referring to buttons or grips by the finger for which they are intended ('the thumb button').

6.2.3 Focus on the Interaction

If recording the task, ensure that the camera is focussed on the product and the tactile interaction taking place rather than the participants face to ensure the exploration is fully documented.

7 Story Telling

Scenarios and storytelling can easily be conducted verbally; however, there are some significant differences between eliciting a verbal story and asking a participant to create a storyboard.

7.1 Points of Interest

7.1.1 Summary Versus Cumulative Viewpoints

When drawing a storyboard, the summary view allows for more information to be added at any stage. As a verbal story is cumulative it is difficult to add more information without breaking the flow of the story. For this reason, the researcher must be clear from the start if they aim to gather any specific information.

7.1.2 Visual Representation

From the researcher's experience, there was little benefit gained from visually representing the verbal stories (post-interview): any attempts made merely replicated the data in a form that was less comprehensible than the original.

8 Conclusion

Engagement and involvement are the two major challenges in a PD process (Sanders 2002). For researchers with little experience of working with people with physical impairments, these challenges are considerably more demanding. It is unlikely that VI participants will have experience of the design process and what it entails; this in itself is engaging for the participants as many of the tasks will be new experiences.

In the experience of the author most participatory tasks can be fairly easily edited to allow for access. However, a prior understanding of the difficulties that may be encountered alongside an early identification of how the output will differ from the initial method is vital for successful research. Generalizable findings include a need for clarity and honesty, a focus on the tactile sense to help engage participants, and a necessary heavier reliance on prior knowledge. Also important to note is the value of finding participants who are invested and interested in the product development. Not only does this mean they are more likely to become engaged with the process but it also means their prior knowledge will be more applicable and relevant. The information presented is intended to be used alongside the existing sources (Gerbe 2000; Henry 2003; Kroll et al. 2007) which give advice on how to best plan and run more traditional UCD research methods.

The participatory process itself has been developed to help researchers gain empathy with users. As with the design of products it is expected that, as the researchers' empathy with the user group grows, they will be better able to successfully design techniques to enhance the product design process through better accessibility, engagement and involvement of VI participants.

References

Brandt E (2006) Designing exploratory design games: a framework for participation in participatory design? In: Proceedings of the 9th conference on participatory design: expanding boundaries in design, Trento, Italy

Gerbe E (2000) Conducting usability research with computer users who are blind or visually impaired. American foundation for the blind. www.afb.org/sectionaspx?SectionID=57&TopicID=167&DocumentID=1718. Accessed 1 July 2013

Henry S (2003) Just ask: integrating accessibility throughout design. http://uiaccess.com/accessucd/resources.html. Accessed 1 July 2013

Kroll T, Barbour R, Harris J (2007) Using focus groups in disability research. Qual Health Res 17(5):690–698

Miao M, Köhlmann W, Schiewe M, Weber G (2009) Tactile paper prototyping with blind subjects. In: Altinsoy ME, Jekosch U, Brewster S (eds) Haptic and audio interaction design. Springer, Berlin Heidelberg

Sanders EBN (2002) From user-centered to participatory design approaches. In: Frascara J (ed) Design and the social sciences: making connections, Taylor and Francis, London

Sanders EBN, Brandt E, Binder T (2010) A framework for organizing the tools and techniques of participatory design. In: Proceedings of the 11th biennial participatory design conference, Sydney, Australia

Embracing Resonance: A Case Study

C. Andrews

Abstract This document provides a short overview of the effects embracing resonance may have on the product design process. It introduces Pullin and Newells definition of resonance and provides examples of where levels of resonance have affected the success of a product. It overviews a case study in which the initial design, a tactile navigation aid for the Blind, was noted as exclusionary for sighted users and documents how recognising the resonance changed both the process and the output. Concluding that early recognition of resonance can enhance the product design process by forcing the designer to consider desirability beyond accessibility needs.

1 Introduction

Innovations created for users with specific impairments have been regularly been documented to create very successful products in the mainstream market (Coleman et al. 2003). This is not always the case. Whilst some products, of which OXO goodgrips are a regularly cited example, explode commercially and do excellently in the mainstream market, others never leave the access market. An example of this is a liquid-level indicator for a mug, as seen on the RNIB website. The difference between these two products is the amount of resonance between the specific market for which they have been designed and the needs of other users.

Pullin and Newell (2007) describe design resonance as: *where the needs of the people who have a particular disability coincide with particular able bodied users in particular contexts.* Hannukainen and Hölttä-Otto (2006) give examples of particular *product usage contexts* (PUC) where resonance might be apparent

C. Andrews (✉)
The National Centre for Product Design and Development Research (PDR),
Cardiff Metropolitan University, Cardiff, UK
e-mail: clandrews@cardiffmet.ac.uk

P. M. Langdon et al. (eds.), *Inclusive Designing*, DOI: 10.1007/978-3-319-05095-9_19,   211
© Springer International Publishing Switzerland 2014

(citing Newell and Gregor 1999) for example: using a laptop whilst standing has resonance with computer users who only have one hand, or communicating in a noisy environment has resonance with deaf or speech impaired people.

OXO goodgrips benefit from resonance; the products were initially conceived to aid a user with arthritis (Coleman et al. 2003). However, there are frequent PUCs in which someone without arthritis would benefit from the innovations, for example, if the user's hands are soapy and wet the large rubber grips would be of benefit. For a liquid-level indicator, on the other hand, there are very rarely PUCs where a sighted person would benefit from the innovation; whilst anybody *can* use this product, few beyond those with a visual impairment need or desire to use it.

Understanding the relationship between the particular physical impairment and the activity limitation and how this resonates with people without this particular impairment is key to predicting whether a product initially conceived for access may have an advantage in the mainstream marketplace; is it usable and desirable by all?

2 Related Literature

After identifying potential resonance the designer must endeavour to fully understand both usage contexts, that of the people *with a particular disability* and that of *able bodied users in particular contexts*. An able bodied user in a particular PUC may be relatively easy to test: simply engineer the context and gather first-hand information, or the designer may already have experience in the particular PUC, which will help identify the problems. Fully understanding someone who lives with these issues on a day-to-day basis will uncover the solutions, but this is more difficult. User-centred design methodologies may help with this process.

User-centred design (UCD) is regularly used to describe a product design process that involves input from the end users at regular stages throughout (Norman 2002); it approaches the design process as an iterative journey with regular meetings with end users to ensure their needs and wants are being answered directly and naturally within the design. This is largely achieved through qualitative research methods such as:

- focus groups;
- observational ethnography;
- usability testing. (Abras et al. 2004).

The guidelines laid out by 'The inclusive design toolkit' (Clarkson et al. 2007) indicates that inclusive/universal design practice fits this mould. However, some practitioners argue that a step beyond this is necessary.

Liz Sanders (2002) lays out participatory design (PD) as a separate methodology to UCD, defining UCD as 'designing **for** users' whereas PD design is 'designing **with** users'. The notable difference is that in traditional UCD methodologies users are not part of the team, but their words and actions are *"spoken for by the researcher"* with a focus on defining needs, wants and desires, whereas in PD the user (or other non-design-based stakeholder) becomes part of the creative team with a focus on experience and gaining empathy (Sanders 2002).

Sanders argues that the traditional UCD research methods allow the designer access to explicit and observable knowledge by watching and talking, but to create true user-based innovations the focus must shift to uncovering tacit knowledge, understanding experience and generating ideas (Sanders 2010).

Lin and Seepersad (2007) propose using 'empathic lead users' as a technique for designers and researchers to develop empathy. They introduce Von Hipple's (1986) idea of lead users as *"customers who push a product to its limits, experience needs prior to the general population, and benefit significantly from having those needs fulfilled"*.

This description could easily be that of a user with an impairment. They suggest that genuine lead users may be hard to find and discuss the possibility of transforming a typical user into a lead user by forcing them to experience the product in a new way (by modifying their environment or the way in which they interact with a product). These exercises are *"designed to break the mold of the customer's thought process and usage pattern"* and so overcome functional fixedness (Hannukainen and Höltta-Otto 2006). An example of this in practice is the Ford 'Third age suit' designed to replicate the problems associated with old age (Steinfeld and Steinfeld 2001). However, whilst a good insight into the surface level needs of specific demographics, it does not give insight into the solutions and coping strategies that these populations utilise on a day-to-day-basis; the activity provides problem-based knowledge rather than solution-based knowledge.

Lilien et al. (2002) suggests that lead users 'may be forced to develop solutions that are novel enough to represent 'breakthroughs' when applied to the target market'.

If this is the case then Lin and Seepersad's problem-based approach does not truly replicate the knowledge gained from lead users. Furthermore, it could be argued that empathic lead users may be overwhelmed with tangential issues and different styles of usage and this may mean that the ability to clearly verbalise what they are experiencing could be difficult; for these reasons empathic lead users will not aid in providing solutions but can aid in identifying problems.

Pullin (2009) agrees that empathy is the key to well designed, accessible products. He introduces a similar concept to lead users, 'extraordinary users' (citing Newell 1997). The term 'extraordinary users' is used to refer to users who have needs beyond that of the average user, which could be due to: an impairment, age (children or the elderly), pregnancy or other similar permanent or temporary conditions and so every one of us will be an 'extraordinary' user at some point in our lifetime (Kroemer 2006). However Pullin takes this a step further and recognises the diversity within each demographic,

each extra-ordinary user should not be considered as representing a specific disability, but
should be considered as an individual person who happens to have a specific disability as
well as a range of other characteristics which are important for defining them as a person.

(Pullin 2009)

He argues that this allows the designer to understand the user more adequately and develop a higher level of empathy. While designing for specific extraordinary users may initially be exclusionary, the advantages are discussed including:

- radical starting points create innovative solutions;
- less constraining than 'fixing' mainstream products;
- encourages simplicity.

He also gives credit to the designer's skill in using the knowledge gained for creating concepts rather than definite solutions. This could be argued to be dependent on the recognition of resonance. If a designer does not recognise how the extraordinary user they are interacting with overlaps with typical users, no effort will be made to make concepts more mainstream.

Hannukainen and Höltta-Otto (2006) show an example of not recognising resonance while arguing for resonance itself, after conducting experiments which conclude that extraordinary users would be valuable in needs assessment in a similar way to lead users. They give various examples of where assistive products have preceded very similar mainstream design. The Memona Plus, a braille note taker, had very similar functionality to the first Palm Pilots released 4 years later. Though they highlight the similarity of needs and how assistive products were ahead of the market, they fail to highlight a potential commercial mistake that the designers of Memona Plus may have made; by not recognising the resonance of needs between VI users and typical users in mobile situations they limited their market to VI users. Often heard is the argument that 'good design' is where designers have considered needs of extraordinary users, but the opposite argument is also regularly true: if the designers of Memona Plus had considered the needs of the typical then perhaps they would have created a more inclusive product, desirable to both VI users and fully sighted users.

2.1 Conclusion

To understand the resonance between an extraordinary user in a typical context and a typical user in an extraordinary context, the designer must understand both sides of the scenario. Whilst understanding how a typical user would react in an extraordinary context is a relatively easy process (for a typical designer!), understanding extraordinary users is more complex. Empathy is key and the use of participatory design methodology alongside the utilisation of 'extraordinary' users as design informants, experts and muses will improve levels of understanding and identification.

3 The Study

3.1 The Case Study

The author was approached by a company to further the development of a conceptual tactile navigation aid designed for visually impaired (VI) users. The concept had originally been designed to be attached to the handle worn by a guide dog. It was quickly identified that whilst the tactile output innovation provided many benefits to target users, the physical design limited use to VI users with guide dogs. Whilst being designed for access, it was exclusionary in execution.

The resonance of the concept was explored further. Multiple scenarios where sighted users would benefit from a tactile output for a pedestrian GPS system were discovered including (but not limited to):

- navigating heavy traffic areas to ensure safety;
- tourists hoping to take in their surroundings while travelling through them;
- users who do not want to publicise that they are in unknown surroundings.

A new brief was formed to take the concept back to the drawing board and rework it with this resonance in mind.

3.2 Previous Work Conducted for the Concept

A review of the work already done was conducted to understand the process and decision-making that led to the conceptual design being harness based.

The methods that had been utilised by the designer fitted well within a UCD framework. The designer gathered explicit and observational knowledge through focus groups and observation of tasks. Some of the participants were selected for further consultation as development proceeded.

Whilst the concept fitted soundly into a UCD approach no participatory or empathy building practices, such as PD or empathic lead users, took place. Problems in the research included:

- lack of recognition of diversity within demographic;
- convenience sampling;
- needs/wants/desires spoken for by the researcher;
- problem identification rather than solution identification;
- observational, hands-off, approach taken.

The approach taken in no way recognised the resonance and potential broader market the product had; rather than utilising VI participants as experts on a grander scale, it restricted itself to being merely an access product. For this reason, further research was conducted to try and understand the solutions that VI people utilise rather than the problems they encounter. This way the solutions can be applied to a broader market.

3.3 Revised Design Process

Whilst VI participants were still utilised in a process that recognised the resonance, their roles changed; VI people became expert informants instead of subjects for specifications to be built around.

Understanding the resonance opened up the brief: suddenly, the specification for research participants changed from 'those who had visual impairments which resulted in trouble navigating' to 'those who are interested in utilising technology to make everyday life easier'; visual impairment no longer became a requirement for the research participant. However, VI users were in an excellent position to be informants on the design due to their greater day-to-day experience of using the tactile sense (also documented by Burnett and Porter 2001).

Summative data gathering techniques throughout the process included:

- focus groups with participatory elements;
- extraordinary users;
- empathic lead user testing.

3.3.1 Focus Groups

The focus group data gained by the original concept designer gave broad information about day-to-day problems for VI people. By recognising the diversity within the demographic and refining the recruitment strategy, data gathered also became more refined. Not only did the participants engage more actively (due to their interest in the subject) but also the knowledge gained was more relevant to the specific product. Three organisations were contacted for recruitment purposes: Action for Blind People, Birmingham (Action); Royal National College of the Blind, Hereford (RNC); and Cardiff Institute for the Blind (CIB). Ideal participants were specified as:

- between 18 and 35;
- actively interested in technology;
- experienced in using mobile devices (phones, navigators and so-on).

Action and RNC were able to gather participants who fitted the specifications, but CIB were unable, probably due to the lack of schools and colleges for VI people in the area. After conducting some initial very open focus groups at both Action and RNC, the participants of the Action focus group proved to have more experience in similar products and so this became the regular focus group. It consisted of five people (three male and two female), aged from 27 to 35 and took place in the home of one of the participants. Focus groups were conducted on a semi-regular basis which helped build up a relationship and a greater depth of empathy and honesty between the design researcher and the participants. The groups were video recorded, transcribed and then coded into themes and opinions.

Some key issues that the focus group discussions uncovered included:

- brand recognition;
- price versus functionality debate;
- comparison to existing products;
- a desire for a mainstream solution.

> *we feel very frustrated that we can't we the same as them, 'they've got a touchscreen why can't we do that?' Apple have turned around and said, well you can now.*
>
> (Participant from Focus Group)

Participatory techniques as discussed by Sanders (2010) were utilised throughout the focus group. Examples include:

- card games to help order functional importance of product features;
- storytelling to gather insight into navigation solutions;
- the use of props to gather feedback on existing handheld electronic devices.

The 'fun' element brought about by the more participatory approach allowed for significantly more engagement and interest in the subject. All three examples produced knowledge that would be difficult to uncover merely through talking alone.

3.3.2 Focus Group Insights

Recognising the diversity of the demographic allowed for a much more in-depth and quality analysis of the market and the users' opinion on it, which helped to understand not only the problems encountered but also what product features help to provide solutions.

Engagement levels were high and enjoyment levels can be assumed to be high as participants attended multiple groups. Participatory methods not only increased engagement but also offered valuable knowledge which would be hard to ascertain through discussion alone.

The focus groups also allowed the researcher to get to know some potential 'extraordinary users'.

3.3.3 Extraordinary Users

From the focus groups, two extraordinary users were selected. Whilst they were both VI they were selected for their expert knowledge of handheld electronic devices, a desire for independence, regular experience in pedestrian navigation tasks and an interest in new products. These selection criteria meant that they fully met the 'potential user' requirements allowing the design to be formed around them. For ethical reasons the participants' names have been removed.

- Participant One is a 35-year-old male living in Birmingham. He states his interest as IT and accessibility and is a self-confessed Apple lover. He studied IT and electronics and regularly travels independently both locally and nationally. He suffers from albinism and is legally registered blind.
- Participant Two is a 30-year-old female also living in Birmingham. She states her interests as music and technology. Whilst owning various Apple products she also particularly likes the brand humanware. She has lived in various places around the UK and so is used to travelling independently. She has been completely blind since birth.

As significant time is spent with extraordinary users the researcher has plenty of opportunity to get to know the participants. Iterative meetings mean that as time goes on the participants get gradually more casual around the researcher resulting in more truthful observations and dialogue.

The extraordinary users were heavily involved in the actual design process. Unlike the original concept where physical entities were developed by the designer than approved by the user, the opposite occurred: physical manifestations of the concept were developed verbally by the extraordinary users and the designer merely helped embody the idea.

Pullin (2009) suggested a flaw in his own argument, that design for extraordinary users may be exclusional for the mainstream, however, the outputs produced by Participants A and B were not at all exclusionary. This could be due to their clear like of mainstream design solutions such as Apple. Whilst radical approaches were discussed regularly, it was the participants who thought on a more practical usable level.

For example, when the concept of a navigation belt was discussed Participant B responded: "*I wouldn't wear that. It's hard enough finding a decent outfit in the first place without having to colour co-ordinate an electronic belt*"

Whilst functionally a belt would have, and has (Pielot 2010) worked very well and very simply, which might make it desirable for a designer to develop, the practicalities and implications of it as an everyday item were not desirable for the extraordinary users.

Foam models and programmable micro-controllers were utilised in the model-making process. The flexibility of these materials allowed the designer to quickly create what the extraordinary users described, ready to be tactilely inspected for development.

3.3.4 Extraordinary User Insights

Extraordinary users provided much of the specification for the physical embodiment of the product, moving it on from concept to realisation.

Spending large amount of time with the extraordinary users allowed a relationship to develop between the design researcher and participants encouraging greater empathy with the participants.

3.3.5 Empathic Lead Users

As discussed in the literature review, empathic lead user testing can be a successful way of understanding problems that may occur in 'extraordinary' user groups. This technique was utilised as a formative study to understand functional issues that may occur with the product as the extraordinary users were some distance away from the design researcher's location.

Three users trialled a prototype. Two male (aged 21 and 28) and one female (25) were recruited through personal contacts using the same specification as the focus groups. All participants had good vision. Users were purposely put in biased scenarios where not only were they blindfolded, but also the path was unmarked and created in an open space. A GPS logger was used to record paths and walking pace, alongside a video recording and short interviews with the participants to gauge their feelings towards the product.

Trials were successful as all participants managed within a short learning time to navigate around the virtual map. However it highlighted problems such as:

- routing algorithm issues;
- command communication issues;
- product rotation to ensure correct compass direction.

Whilst some of these issues were technical, design issues, such as product rotation were taken back to the extraordinary users who had experience and insight into potential solutions.

3.3.6 Preliminary Results

The new solution is being designed to be functional and desirable. The current concept plays on current trends in the mobile device market in creating a device which works alongside smartphone apps and it can be utilised by users with any level of vision. The product preserves, and expands on, the innovative tactile output developed in the original whilst not being exclusionary for sighted users. It is currently in the testing stages with participants who were not involved with the development process; attributes being tested include: functionality, usability and desirability with a variety of potential users.

Preliminary results seem positive indicating that though journey time has slightly increased with the unfamiliar tactile device, in comparison to a standard visual interface, users enjoy its innovative nature and ability to take in their surroundings.

4 Conclusion

Early identification of design resonance can significantly change both the process and outcome involved in the design of a product. Whilst many product ideas are originally conceived as solutions to issues people with physical and mental impairments face on a day-to-day basis the recognition of a broader audience will aid commercial value. Recognition of resonance is a solution-based design approach as it utilises ideas, solutions and coping strategies of the (relatively) few who deal with a problem everyday and makes them accessible to many who may deal with similar issues on a more occasional basis.

On the other hand, it may mean that the designer recognises that a product does not have enough resonance to be practical for a mainstream market and so allows them to scope their research and refine their product for a specific target market.

In either scenario, the product can be analysed and developed in a way appropriate to the final market. Products which were initially conceived to help a small group of people may develop into truly inclusive products that help many.

References

Abras C, Maloney-Krichmar D, Preece J (2004) User-centered design. In: Bainbridge W (ed) Encyclopedia of human-computer interaction, vol 37. Sage, Thousand Oaks, pp 445–56

Burnett GE, Mark Porter J (2001) Ubiquitous computing within cars: designing controls for non-visual use. Int J Hum Comput Stud 55(4):521–531

Clarkson PJ, Coleman R, Hosking I, Waller S (2007) Inclusive design toolkit. Engineering design centre, University of Cambridge, UK

Coleman R, Lebbon C, Clarkson PJ, Keates S (eds) (2003) From margins to mainstream. Inclusive design, design for the whole population. Springer, London

Hannukainen P, Hölttä-Otto K (2006) Identifying customer needs - disabled persons as lead users. In: Proceedings of the ASME IDETC design theory and methodology conference, Philadelphia, US

Kroemer K (2006) Extra Ordinary Ergonomics: how to accommodate small and big persons, the disabled and elderly, expectant mothers, and children. CRC Press, Florida, USA

Lilien GL, Morrison PD, Searls K, Sonnack M, von Hippel E (2002) Performance assessment of the lead user idea-generation process for new product development. Manage Sci 48:1042–1059

Lin J, Seepersad CC (2007) Empathic lead users: the effects of extraordinary user experiences on customer needs analysis and product redesign. In: Proceedings of the ASME DETC design theory and methodology conference, Las Vegas, US

Newell AF, Gregor P (1997) Human computer interfaces for people with disabilities. In: Helander M, Landauer TK, Prabhu P (eds) Handbook of Human-Computer Interaction. Elsevier Science, Amsterdam, The Netherlands

Newell AF, Gregor P (1999) Extra-ordinary human-machine interaction: what can be learned from people with disabilities? Cogn Technol Work 1:78–85

Norman DA (2002) The design of everyday things. Basic books, New York

Pielot M, Boll S (2010) Tactile wayfinder: comparison of tactile waypoint navigation with commercial pedestrian navigation systems. In: Proceedings of the pervasive computing, Helsinki, Finland

Pullin G, Newell A (2007) Focussing on extra-ordinary users. In: Proceedings of the universal access in human computer interaction, Beijing, P.R. China

Pullin, G (2009) Design Meets Disability. MIT Press Books, USA

Sanders EBN (2002) From user-centered to participatory design approaches. In: Frascara J (ed) Design and the social sciences: making connections. CRC Press, London

Sanders EBN, Brandt E, Binder T (2010) A framework for organizing the tools and techniques of participatory design. In: Proceedings of the participatory design conference, Sydney, Australia

Steinfeld A, Steinfeld E (2001) Universal design in automobile design. In: Preiser WFE, Ostroff E (eds) Universal design handbook. McGraw Hill Professional, New York

Von Hippel E (1986) Lead users: a source of novel product concepts. Manage Sci 32(7):791–805

Can We Work Together? On the Inclusion of Blind People in UML Model-Based Tasks

L. Luque, E. S. Veriscimo, G. C. Pereira and L. V. L. Filgueiras

Abstract Every person, regardless of age, gender, race, or any other circumstances, has the right to fully participate in society, having a life as normal as possible. Unfortunately, this does not correspond to reality. Aiming to change this scenario, there has been a growing effort to promote social inclusion worldwide. When considering people with disabilities, this can be noted by the establishment of laws and public policies that seek to ensure their rights. Although the field of computation plays an important role in this context, allowing the development of technologies that promote more independence and autonomy for people with disabilities, many tools and technologies used in this field are still inaccessible, which makes difficult the inclusion in computer education programs and in the industry. Among these tools are various graphical notations, such as the Unified Modeling Language-UML, which are inaccessible to blind people. This graphical notation is extensively used to simplify, understand and document different aspects of object-oriented software systems. Although some studies in the literature propose solutions that enable the access to information present in UML models for the visually impaired, there are no studies that formally analyzed the accessibility of UML modeling tools being used in computer education programs and in the industry. Moreover, visually impaired students and professionals with whom the authors of this study had contact reported a difficulty in finding solutions that improve their accessibility. In this context, this paper aims to assess the availability and accessibility of solutions for the inclusion of blind people in activities focused on the creation and editing of UML diagrams, allowing the inclusion of these people in computer education programs and in the software development

L. Luque (✉) · E. S. Veriscimo · G. C. Pereira
Department of System Analysis and Development, São Paulo State Technological College
(FATEC), Mogi das Cruzes, Brazil
e-mail: leandro.luque@fatec.sp.gov.br

L. V. L. Filgueiras
Polytechnic School of Engineering, University of São Paulo, São Paulo, Brazil

P. M. Langdon et al. (eds.), *Inclusive Designing*, DOI: 10.1007/978-3-319-05095-9_20,
© Springer International Publishing Switzerland 2014

industry. The results indicate that the UML modeling tools used in the industry are not accessible to blind developers and the alternative accessible tools are not appropriate for use in the industry.

1 Introduction

Every person, regardless of age, gender, race or any other circumstances, has the right to participate fully in society, having as normal a life as possible (UN 1948). Unfortunately, this does not correspond to reality (Marlier 2007).

Aiming to change this scenario, there has been a growing effort to promote social inclusion worldwide. In the case of people with disabilities, this can be seen by the establishment of laws and public policies that seek to ensure their rights (Brazil 2004; ADA Amendments Act 2008; UK 2010).

Although the field of computing plays an important role in this context, allowing the development of technologies that promote more independence and autonomy for people with disabilities, many tools and technologies used in this field are still inaccessible, which makes inclusion in computer education programs and in industry difficult. Much information about these difficulties can be found at the National Science Foundation (NSF) Access Computing Knowledge Base.

Among these tools are various graphical notations, such as entity-relationship and data flow diagrams, which are inaccessible to blind people (Cohen et al. 2006; Metatla et al. 2008). These are extensively used to simplify, understand and document different aspects not only of software systems but also of other kinds of systems.

The Unified Modeling Language (UML) is one of the most used notations in software systems development (Dobing and Parsons 2006), widely adopted in industry and part of core units of the reference curricula for the area (Table 1).

Although some studies in the literature propose software solutions that enable access for visually impaired people to information present in UML models, it was not clear whether or not any of these solutions were really implemented and available for download. Also, we have not found studies that formally analysed the accessibility of UML modeling tools now being used in computer education programs and in industry. Moreover, visually impaired students and professionals with whom we have contact reported difficulties in finding solutions that improve their accessibility.

In this context, this paper aims to assess the availability and accessibility of solutions for the inclusion of blind people in activities focused on the creation and editing of UML diagrams, facilitating their inclusion in computer education programs and in the software development industry.

The remainder of this paper is organised as follows. Section 2 provides a brief overview of UML, its usage in classroom and in industry and of how blind users usually interact with computers. Also, this section presents the related work on

Table 1 UML in the reference curricula for the field of computing

Reference curricula for	Unit/course
Computer science (ACM/IEEE 2008a)	Data modeling
Computer engineering (IEEE/ACM 2004a)	Data modeling
Information tech. (ACM/IEEE 2008b)	Data modeling and integrative coding
Information systems (ACM/AIS 2010)	Systems analysis and design
Software engineering (IEEE/ACM 2004b; IEEE/ACM 2009)	Introduction to software engineering and computing

solutions to make UML diagrams accessible to visually impaired people. In Sect. 3, we present the methodology followed in this study. The results are presented in Sect. 4 and discussed in Sect. 5.

2 Background and Initial Considerations

2.1 UML in Classroom and in Industry

UML is a standard notation that can be used to understand, specify and document software systems (Booch et al. 2006). It has 14 different types of diagrams, categorised into structural and behavioural, and it is used in different development phases and disciplines, such as business modeling, requirements, design and so on.

Use case and class diagrams are the most used UML notations (Dobing and Parsons 2006). The former represents a system from an external point of view and includes elements such as: actors, use cases, associations, generalisations and dependencies (Fig. 1, left). The latter allows the representation of the static structure of a system and contains elements such as classes, attributes/fields, methods/operations, associations, aggregations, compositions, generalisations, dependencies and so on (Fig. 1, right).

In computer-related courses, many lecturers use UML models to teach students object-oriented analysis and programming, as it is part of the reference curricula for this area (IEEE/ACM 2009). Usually, it is not possible for blind students to skip UML classes or prove their knowledge with alternative tasks (Müller 2012).

To participate in UML model-based tasks in classroom or in industry, a blind person should be able to:

- create models in a format that can be transmitted to the rest of the team/teachers/ colleagues;
- have access to models created by other people and be able to use them to carry out their activities (e.g. the development of a component from its specification);
- change the models created by him/herself and others.

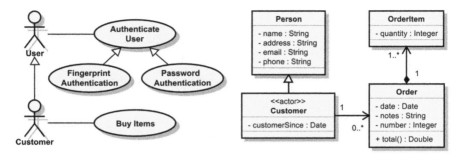

Fig. 1 Example of a use case diagram (on the *left*) and example of a class diagram (on the *right*)

These requirements are related to each other in the sense that it is desirable that both sighted and blind people use the same tool to perform them. In the academic context, it is easier to produce cultural changes to include blind students, such as using alternative tools or changing sighted students' procedures. But these changes are less likely to be adopted in enterprise environments, where projects often have tight deadlines and the use of productivity tools ends up being essential.

2.2 How Blind Users Use Computers

Blind users access information through other senses, especially hearing and touch. When considering computers, touch involves the use of tactile displays and printers, while hearing involves auditory interfaces.

A common solution for auditory interfaces is the use of screen readers. They synthesise texts and graphical components found in other software interfaces. According to a survey by WebAIM (2012), the five most used screen readers are JAWS, NonVisual Desktop Access—(NVDA), VoiceOver, System Access or System Access To Go (SA or SAToGo) and Window-Eyes. In Brazil, together with JAWS, Virtual Vision and Dosvox/Webvox (both developed in Brazil) rank as the most used.

When using graphical user interfaces (GUI), such as those used in modeling tools, blind users face several barriers (Donker et al. 2002):

- pixel barrier: the pixels that comprise the interface are not readable by screen readers;
- mouse barrier: blind users cannot handle the mouse in an effective way—data input is usually performed by means of the keyboard (menus and shortcuts) and voice commands;
- graphics barrier: transforming images to other media always means a loss of information;
- layout barrier: occurs when important information is conveyed by the location of images on the screen.

2.3 Related Work

Although the literature on accessibility of different types of diagrams and graphs is extensive, there are a few studies which seek to make UML diagrams accessible to visually impaired people by means of different strategies. The strategies proposed in these studies involve the use of haptic solutions, tables with information, communication with existing tools using a common format, such as XMI (XML Metadata Interchange), among others.

King et al. (2004) presents a software called TeDUB that allows to be imported diagrams from different sources. When considering UML diagrams, it allows the import of XMI files, generated by almost all UML CASE (Computer-Aided Software Engineering) tools.

Brookshire (2006) used tactile cards to teach class diagrams to one blind student. The author says that the quality of the diagram created by the student was almost the same as the one created by sighted students.

Silva et al. (2010) propose the use of a spreadsheet for creating and reading UML diagrams. As an example, a use case diagram can be represented as follows: the first column lists actors, the second column lists use cases, the third column contains the relationship types and the fourth column the other actors/use cases for relationships.

Loitsch and Weber (2012) propose an audio-tactile screen explorer, called HyperReader, that allows blind people to access the information present in UML sequence diagrams. This screen explorer can be integrated with Microsoft Visio to make it possible for sighted people to collaborate.

The authors evaluated the understanding of sequence diagrams and the work-load associated with reading graphical sequence diagrams using their solution in comparison to a verbalised representation.

The results showed that the latter presented better results than HyperReader in almost all tasks performed, leading the authors to conclude that a representation combining UML diagrams and a verbalised solution is meaningful, particularly to obtain information for specific tasks quickly.

Müller (2012) reports on her experience of teaching UML to two blind students using tactile class and sequence diagrams printed with Emprint of ViewPlus and described textually, through a table structure (class) and a list of messages (sequence). No experiment was conducted to compare this approach with existing ones.

Pansanato et al. (2012) suggest the use of automatic conversion of XMI specifications into a table format.

Additionally to studies on the accessibility of UML diagrams, there are some general approaches aiming to make any kind of diagram accessible. Goncu (2009), for example, proposes an architecture to automatically generate accessible diagrams from a standard input format (e.g. SVG or image) through the extraction of syntax, semantics and pragmatics from the diagram and the combination of them with a user profile and contextual information.

Another approach is GraSSML (Fredj and Duce 2007), a solution based on two languages: one to represent the structure of a diagram (ZineML) and another to represent the semantic intent behind the diagram (MyLanguage). Through the representation of a diagram in these two languages, it would be possible to convert the diagram into any accessible format.

3 Evaluation and Research Method

To verify the availability of the aforementioned software solutions, we performed searches on code repositories (CodePlex, GitHub, Google Code and SourceForge) and on Google.

After that, to evaluate the accessibility of existing UML modeling tools, we first conducted a survey on the use of these tools by software development companies in the State of Sao Paulo/Brazil, where more than a third of all Brazilian information technology companies are based (MTE 2011).

The survey was sent to fifty randomly selected companies and twenty of them responded. The results are shown in Table 2. All the tools mentioned by respondents were selected for evaluation.

We decided to conduct the evaluation considering only use case and class diagrams since they are the two most widely used UML diagrams. As will be shown later, this is not restrictive, because the accessibility problems found are related to general aspects that are independent of the diagram type.

We chose the screen reader JAWS to conduct the evaluation since it is frequently used by visually impaired users.

The evaluations were conducted as usability experiments, with three users. One of them is blind and two are sighted. The blind user has been using JAWS daily for 4 years to use computers and develop software. One of the sighted users is an instructor in human–computer interaction and has been using JAWS to some degree for about 3 years. He is a frequent speaker on assistive technologies for visually impaired users. The other user received training in this type of tool.

Users were given a narrative, with all the information necessary to carry on the evaluation, related to various activities summarised in Fig. 2.

4 Results

4.1 Availability of Proposed Solutions

With only one exception, none of the proposed software solutions was found available for download. The exception was TeDUB (King et al. 2004), see Sect. 2.3, that allows users to explore UML diagrams using a keyboard or a

Table 2 UML modeling tools used by software development companies in Sao Paulo/Brazil

UML modeling tool	Number of companies	Fraction of companies (%)
ArgoUML v0.34	1	5
Astah professional 6.5.1	3	15
Enterprise architect ultimate 9.0	5	25
Jude community 5.4	1	5
Smartdraw 2013	1	5
Rational modeler 7.5.0	2	10
Visio 2013	3	15
Do not use	4	20
	20	100

1. Create a use case diagram named "e-Commerce". Add three actors: "Customer" (A1), "User" (A2) and "Manager" (A3). Add six use cases: "Authenticate User" (UC1), "Authenticate Password" (UC2), "Authenticate Fingerprint" (UC3), "Buy Product" (UC4), "Send Invoice" (UC5) and "Generate Sales Report" (UC6). Create a generalization relationship between use cases: UC2 > UC1 and UC3 > UC1. Create a dependency inclusion between UC5 and UC4. Associate: actor A1 to the use case UC4, actor A2 to the use case UC1, and actor A3 to the use case UC6. Finally, establish a generalization between: actors A2 > A1 and A3 > A1.

2. Open an existing use case diagram and answer the following questions:
   a. Who are the actors in the diagram?
   b. What are the use cases of the diagram?
   c. What are the relationships among the elements of the diagram?

3. Change the diagram opened in activity 2, adding an actor and a use case and linking them to existing elements.

4. Create a class diagram and add five classes to it: Person (C1), Customer (C2), Order (C3), Product (C4) and OrderItem (C5). Add the following attributes/methods to the classes:
   Person: name, address, phone, e-mail;
   Customer: code, customerSince;
   Order: date / addProduct, removeProduct, total;
   Product: code, name, description;
   OrderItem: quantity.

5. Open an existing class diagram and answer the following questions:
   a. What are the classes of the diagram?
   b. What are the attributes and methods of each class?
   c. How do the classes relate to each other?

6. Change the diagram opened in activity 5, adding a new class and linking it to an existing class.

Fig. 2 Narrative to be used in the evaluation

joystick. The tool enables users to open UML diagrams in the XMI format, without allowing editing of content, which limits the full integration of blind and sighted people.

During the search process, we found a few alternative tools (not listed in the literature) that allow the generation of some UML diagrams from textual

Table 3 Web-based tools to generate UML diagrams from textual descriptions

Web tool	Supported diagrams
Diagrammr	Interaction
Umple	Class and state machine
WebSequenceDiagrams	Sequence
yUML	Class, use case and activity

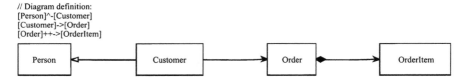

Fig. 3 yUML diagram and the respective textual description generated by Astah for the class diagram represented in Fig. 1 (*right*)

descriptions. Since they are web-based tools that use normal web forms, they are accessible for blind users (Table 3).

Due to a recent integration with Astah Professional, diagrams created with Astah can be exported to yUML. However, the solution is still limited, creating only a one-way communication channel, since it is still not possible to import files generated by yUML into Astah.

Furthermore, the exportation of class diagrams does not work for attributes/ fields and methods/operations (Fig. 3), and only three types of diagrams are supported by yUML.

There are also standalone tools that support the generation of some UML diagrams through textual descriptions, such as: TextUML Toolkit, MetaUML, PlantUML, EasyUML editor, seqdiag, among others. As the web equivalents, they suffer from lack of integration with existing UML CASE tools, as well as the support of only a few diagram types (mainly sequence and class) and some of them are also inaccessible.

4.2 Accessibility Evaluation

Regarding activities 1, 3, 4 and 6, none of the participants was able to create or edit use case or class diagrams. The most inaccessible interfaces were of Astah, Jude and Smartdraw, for which none of the menu options were read by JAWS, making it impossible to know in which option or context the user was. It is important to note that all computers used to conduct the evaluation had the Java Access Bridge installed, so that Java Swing tools, such as Astah and Jude, should have been accessible.

Table 4 Summary of results of the accessibility evaluation

UML modeling tool	Result by task					
	1	2	3	4	5	6
ArgoUML v0.34	No	No	No	No	No	No
Astah professional 6.5.1	No	No	No	No	No	No
Enterprise architect ultimate 9.0	No	No	No	No	No	No
Jude community 5.4	No	No	No	No	No	No
Smartdraw 2013	No	No	No	No	No	No
Rational rose/Modeler 7.5.0	No	Partially	No	No	Partially	No
Visio 2013	No	No	No	No	No	No

Although for ArgoUML, Enterprise Architect, Visio and Rational Modeler menus were correctly read, it was not possible to create or edit elements, because the functionality for that is strongly linked with the use of the mouse.

As for activities 2 and 4, only Rational Modeler allowed the reading of some diagram elements—through the use of the 'window virtualisation' feature in JAWS. However, information about the relationships could not be identified correctly, preventing a complete understanding of the diagrams.

Trying to find shortcuts to complete the specified activities, participants made a systematic search for shortcut combinations using all characters individually and combined with the modifiers Alt, Control and Shift. No useful shortcut was found. The two most experienced users with screen readers even tried to use the JAWS cursor feature, also without success.

Table 4 summarises the results of the accessibility evaluation.

5 Discussion

The inclusion of blind people in computer education programs and in the software development industry generally involves the use of graphical notations, such as UML diagrams.

However, UML CASE tools used in universities and in the industry are not accessible and do not allow collaborative work with sighted developers. As an alternative to these, there are haptic (e.g. Brookshire 2006; Müller 2012; Loitsch and Weber 2012) and textual solutions (e.g. yUML and PlantUML).

Both are less likely to be used in companies, where projects often have tight deadlines and the use of productivity tools ends up being essential.

Regarding the academic context, where the culture can be more easily changed, these tools can be used, but they do not support all UML diagrams. Also, as haptic solutions involve the knowledge of Braille, they are not suitable for all blind students, since not all of them know Braille.

The solutions that are more likely to be used in industry are not available to the public and seem to be restricted to the research context. The only available tool found, TeDUB, only allows the reading of some diagrams without editing the features.

An alternative, specifically to class and sequence diagrams, is the generation of code through UML CASE tools. However, in addition to not allowing the differentiation of certain details (such as simple association/aggregation/composition), the small number of languages that can have a code generated excludes many developers.

This study showed that not all UML diagrams are accessible and there is a challenging scenario for the inclusion of blind developers in the industry, making further researches important in this area.

Acknowledgments The authors wish to thank Alex Braha Stoll, Andrea Massumi and Valdite Fuga for carefully reading this article. This work was funded by São Paulo State Technological College (FATEC Mogi das Cruzes).

References

ACM/AIS (2010) IS 2010. Curriculum guidelines for undergraduate degree programs in Information Systems

ACM/IEEE (2008a) Computer science curriculum 2008: an interim revision of CS 2001. Report from the interim review task force

ACM/IEEE (2008b) Information technology 2008. Curriculum guidelines for undergraduate degree programs in information technology

ADA Amendments Act (2008) United States department of labour, Washington, DC, US

Booch G, Rumbaugh J, Jacobson I (2006) UML. Elsevier, Brasil

Brazil (2004) Executive Act no. 5.296/2004. On the rights of people with disabilities

Brookshire RG (2006) Teaching UML database modeling to visually impaired students. Issues Inf Syst 7:98–101

Cohen RF, Haven V, Lanzoni JA, Meacham A, Skaff J et al (2006) Using an audio interface to assist users who are visually impaired with steering tasks. In: Proceedings of the 8th international ACM SIGACCESS conference on computers and accessibility, Portland, OR, US

Dobing B, Parsons J (2006) How UML is used? Commun ACM 49(5):109–113

Donker H, Klante P, Gorny P (2002) The design of auditory user interfaces for blind users. In: Proceedings of the 2nd Nordic conference on human-computer interaction, Arhus, Denmark

Fredj ZB, Duce DA (2007) GraSSML: accessible smart schematic diagrams for all. Univ Access Inf Soc 6(3):233–247

Goncu C (2009) Generation of accessible diagrams by semantics preserving adaptation. ACM SIGACCESS Access Comput 93:49–74

IEEE/ACM (2004a) Computer engineering 2004. Curriculum guidelines for undergraduate degree programs in computer engineering

IEEE/ACM (2004b) Software engineering 2004. Curriculum guidelines for undergraduate degree programs in software engineering

IEEE/ACM (2009) Graduate software engineering 2009. Curriculum guidelines for graduate degree programs in software engineering

King A, Blenkhorn P, Crombie D, Dijkstra S, Evans G, Wood J (2004) Presenting UML software engineering diagrams to blind people. In: Computers helping people with special needs, Paris, France

Loitsch C, Weber G (2012) Viable haptic UML for blind people. In: Proceedings of the 14th international conference on computers helping people with special needs, Linz, Austria

Marlier E (ed) (2007) The EU and the social inclusion: facing the challenges. The Policy Press, Bristol

Metatla O, Bryan-Kinns N, Stockman T (2008) Constructing relational diagrams in audio: the multiple perspective hierarchical approach. In: Proceedings of the 10th international ACM SIGACCESS conference on computers and accessibility, Halifax, Nova Scotia, Canada

Müller K (2012) How to make unified modeling language diagrams accessible for blind students. In: Proceedings of the 14th international conference on computers helping people with special needs, Linz, Austria

MTE (2011) Ministério do Trabalho e Emprego. www.investe.sp.gov.br/setores/tecnologia-informacao. Accessed 29 Oct 2013

Pansanato LTE, Bandeira ALM, Santos LG, Pereira DP (2012) Projeto D4ALL: acesso e manipulação de diagramas por pessoas com deficiência visual. In: Proceedings of the 11th Brazilian symposium on human factors in computing systems, Cuiaba, Brazil

Silva CE, Pansanato LT, Fabri JA (2010) Ensinando diagramas UML para estudantes cegos. In: Proceedings of the 36th latin american informatics conference, Asunción, Paraguay

UK (2010) Equality Act 2010. www.legislation.gov.uk/ukpga/2010/15/contents. Accessed 29 Oct 2013

UN (1948) The universal declaration of human rights. Resolution adopted by the General Assembly 10/12. United Nations, New York, US

WebAIM (2012) Screen reader user survey #4. http://webaim.org/projects/screenreadersurvey4/. Accessed 29 Oct 2013

Co-design in China: Implications for Users, Designers and Researchers

S. Yuan and H. Dong

1 Introduction

A significant shift in design has led to increased emphasis on and involvement of the user as a critical resource and a source of inspiration for the designer (McDonagh 2008). Terms like participatory design and co-design describe user-involved design processes (Spinuzzi 2005; Sanders and Stappers 2008) where users or customers are no longer passive subjects, but active co-designers or 'partial' employees (Bitner et al. 1997; Sanders and Stappers 2008; Kaasinen et al. 2012), who are 'experts of their experience' (Sleeswijk et al. 2005) and special for their 'tacit knowledge' (Spinuzzi 2005).

Participatory design originated in 1970s in Norway, when computer professionals and union leaders tried to give workers some influence on the shape and scope of new technologies introduced into the workplace (Winograd 1996; Spinuzzi 2005). According to Muller and Druin (2007), participatory design began in an explicitly political context, followed by the consideration of additional social justice issues, such as inclusive design, women's needs, cultural sensitivity, etc. Nowadays, participatory design processes are being applied to urban design, planning, geography, as well as to the fields of industrial and information technology (Sanoff 2007).

Co-design, as the process of designing with people who will use or deliver a product or service, engages non-designers by asking, listening, learning, communicating and creating solutions collaboratively (Design Council 2013). As a format of participatory design, co-design is more frequently used in product and

S. Yuan
Tongji University, Shanghai, China
e-mail: yuanshu66@gmail.com

H. Dong (✉)
Inclusive Design Research Centre, College of Design and Innovation,
Tongji University, Shanghai, China
e-mail: donghuahao@gmail.com

P. M. Langdon et al. (eds.), *Inclusive Designing*, DOI: 10.1007/978-3-319-05095-9_21,
© Springer International Publishing Switzerland 2014

service design. In this paper, 'co-design' is used to refer to the user-involved design process.

As co-design is a relatively new concept in China, this study aims to explore its implications for users, designers and researchers. The research questions are:

- How do users participate in the co-design process?
- What is the designer's role in the co-design process?
- How can researchers better support co-design activities?

Two studies were conducted in Shanghai during 2012–2013. In the first, a half-day co-design workshop was organised for professional designers and diverse users, adopting the short format of the Inclusive Design Challenge Workshops Model (Cassim and Dong 2007, 2013). The second study was a 6-week long project with undergraduate students from Tongji University, using the 'Fixperts' project brief (available on http://www.fixperts.org) where the design students ('fixperts') were asked to collaborate closely with a research partner ('fixpartner') to solve a practical problem.

2 Methods

2.1 Co-design Workshop

In the co-design workshop, five professional industrial designers (each with more than 5 years of professional design experience) and five diverse users (with different ages and abilities) participated in a half-day co-design workshop in the Design Factory of Tongji University, Shanghai. Experienced designers were selected to reflect the real-world design practice, and people with severe disabilities acted as a stimulus for design innovation (Cassim and Dong 2013).

Before the workshop, the designers and users were informed about the workshop process and were asked to prepare self-introduction materials and bring along their favourite and least favourite design. In order to establish mutual trust, pre-workshop meetings were arranged between the researchers and the users, and four users were visited at home. As users were not necessarily familiar with the concept of 'design', the researchers explained the meaning of design to them, and helped them to find design examples surrounding them.

In the workshop, pre- and post-questionnaires were distributed to the participants. The pre-questionnaires for the designers and the users had similar questions (e.g. 'What are your criteria for judging good design and bad design?', 'What role do you think you can play in the design process?'). Immediately after the workshop, post-questionnaires were handed out, which aimed to understand the participants' existing knowledge and practice of design and to find the impact of the workshop on participants. The designers and the users were mixed into three groups to do a co-design task. During the co-design session, researchers were allocated to each group to observe the interactions between the designers and the users. They objectively recorded what happened in the co-design process. Follow-up interviews were arranged with the designers and the users for reflection.

2.2 'Fixperts' Project

The second study was embedded in the 'User research' module (taught by the authors) for undergraduate design students at Tongji University. The 'Fixperts' project brief was adopted where students (i.e. fixperts) were required to closely collaborate with a 'fixpartner' (a person from the public). In total, 72 third-year undergraduate design students participated in the project, their backgrounds including product design, visual communication design, environmental design and digital media design. They self-formed 14 groups; each with four to six students. Half of the groups found young students as their 'fixpartners', and the other half had a diverse range of people as their 'fixpartners'. The students were asked to record their design process during the 6-week project period, and make a 3-min film at the end of the project.

Common user research methods (i.e. observation, interview, persona, scenario etc.) were introduced in lectures. EMPATHY was emphasised as young designers might be impatient when they cooperate with users (Mcdonagh 2008; Ho et al. 2011). The first author (a PhD researcher) shadowed eight groups of students when they met their fixpartners, with the purpose of observing the interaction between the fixperts and the fixpartners in a natural setting. Towards the end of the project, the students were asked to fill out a four-page A3 questionnaire in the class. This questionnaire aimed to assess whether the students had empathy with their fix-partners, how the co-design process was carried out and what was the students' understanding of 'designing with people'. The structure of the questionnaire is shown in Fig. 1. Both drawings and texts were encouraged for answering the questions, as drawings are a common way for designers to communicate thinking.

The six groups that were not shadowed were interviewed by the Ph.D researcher after the project, to obtain additional information not captured by the questionnaire.

3 Results

In this section, the results of the co-design workshop and the fixpert project are reported.

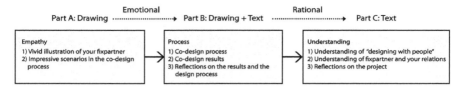

Fig. 1 The structure of the 'fixperts' questionnaire

3.1 Co-design Workshop

The comparison of the pre- and post-questionnaires of the users suggests that they developed an understanding of design from the workshop. However, the designers' understanding of design and their criteria for judging good/bad design did not seem to change much after the workshop, although 'inclusion' started to appear in the post-questionnaire, suggesting the co-design workshop had some impact on the designers.

The follow-up interviews suggest that although all the users appreciated their participation in the workshop, especially the opportunity to get to know designers and the design process, some of them were not comfortable in the co-design session as they felt that designers were dominating the process and were proposing 'nice-looking but not practical' solutions to the problem (Dong and Yuan 2013). The observation by the researchers revealed that in one group, the designers tended to talk to other designers more, and they used jargon (e.g. 'function' and 'ratios') which made it difficult for the user to join in the discussion. One designer even interrupted the user once and began to talk about her own ideas. Sometimes the designers tried to involve the user in the conversation by asking 'what's your opinion on this, Mr. Xie?' However, such sudden and general questions often made the user tongue-tied. On the other side, the users' ideas did not seem to inspire the designers, which made the co-design process not so exciting for the designers. In another group, when confronted by a different opinion from the users, the designers simply made a compromise rather than investigating why the users said that (Dong and Yuan 2013). It was observed that the designers often ignored the users' concept when drawing concept sketches. In the follow-up interview, one designer in this group complained that the design 'taste' of the user was not good, while the user thought this designer made design that 'only works on paper, but not in practice'.

In the follow-up interview, all the five designers admitted that designers needed to get to know users. However, two of them mentioned that face-to-face contact with users might not be the optimal way to know users. One preferred traditional focus groups rather than negotiating with the 'stubborn user'. The other designer euphemistically expressed the idea that 'there was no need to do user research in person if designers were already familiar with the design objective'. These indirectly show designers' suspicion of co-design.

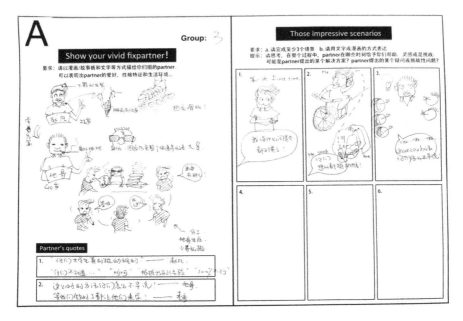

Fig. 2 One student's drawing about their group fixpartners

3.2 'Fixperts' Project

As close collaboration with the fixpartner was required for the Fixpert project, students made great efforts and each group met their fixparters five to seven times during the 6 weeks. All the students showed politeness and respect to their fixpartners. In order to save time for their fixpartners, they arranged meetings at their fixpartner's study/work place or a café nearby. Although difficulties in communication and understanding were observed between the students and their fixpartners, no interruption ever happened. In one group, after the students explained to their fixpartner (a cleaner at the Design Factory) that they wanted to help her solve a practical problem, the cleaner, having no knowledge of what designers do, thought for a while and asked one male student in the group to help strap the mop for her. When this happened, the students patiently explained what designers do and in what way they could help her, and tried their best to inspire the cleaner to express difficulties in her daily life.

Drawings from Part A of the questionnaire suggest that students had empathy with their fixpartners. They drew interesting cartoons and storyboards to describe their fixpartners and their problems. Figure 2 shows one student's drawing about their fixpartners (two brothers who deliver express post to the University) and their working environment. The drawing illustrates the problem: massive amounts of post and numerous inquires from the students often made the younger brother really frustrated.

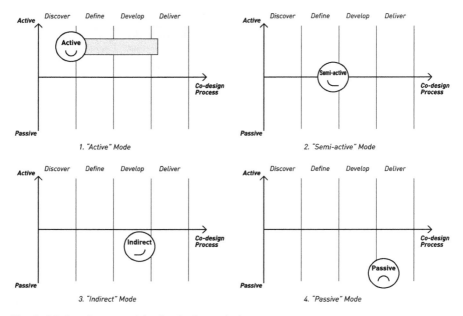

Fig. 3 Modes of user participation in the co-design process

The results of Part B and Part C were analysed and user participatory models are summarised in Fig. 3. The 4D model (Design Council 2005) was adopted to describe the generic design process, i.e. discover, define, develop and deliver.

There are four modes of user participation in the Fixperts project, namely, active, semi-active, indirect and passive (Fig. 3). In the active mode, users are conversable and are completely immersed in the co-design process, and their active involvement lasts till the early 'deliver' stage. Four out of the fourteen groups fixpartners were in this active mode, and three of them were students with a design background (e.g. architectural design). It is worth noting that one 'active' fixpartner, Mr. Zhao (the younger man who delivers express posts to university students, shown in Fig. 2), has no design background. He is of similar age to the students and they communicated like friends. Although Zhao did not receive much education, he is very proud of his rich life experiences. For example, in Fig. 2, the student quoted Zhao's words 'you university students sometimes are too naïve'. Although Zhao did not directly contribute to the final design solution, his active participation indeed gave the students inspirations and made the whole design process more smooth and enjoyable.

In the 'indirect' mode, users neither show much activity nor tell the designers their intentions or expectations. But they are willing to have a conversation with the designers, which may inspire design solutions. For instance, in one group, the fixpartner was a fitness instructor who had a busy lifestyle, but she had made an effort to meet the students a few times. Through open conversations, the students found that her slight blue–green colour blindness had caused awkwardness in her

Fig. 4 The keyring colour references: *green leaf* and *blue dew drops*

daily life. They decided to find a solution to tackle this specific problem. The students designed a series of colour tests, showing different combinations of blue and green cards to the fixpartner, until they found a pair that she could easily differentiate and use as a colour reference. During the breaks in the tests, the students chatted with the fixpartner to get an insight into her life style and preferences. By chance they found she always carried two key rings (one for home and the other for work) with her, and this inspired the students to apply the blue and green colours in the key rings as her discrete colour reference (Fig. 4). In this case, the user did not know her problem or expectation, but her words inspired the designers to identify the problem and find a solution which suits her very well.

In the passive mode, users participate less in the idea generation phases. They give feedback to design outcomes, often in the latter 'develop' phase or the early 'deliver' phase when prototypes are available. In this mode, the interaction between users and designers is rather passive and their relations are less close than in the other modes.

The four modes summarise the pattern of user participation; in some cases, different modes coexist and there may be transitions from one to another. For example, in one group, the fixpartner converted from the passive mode to the active mode in the latter 'develop' phase, which changed the dynamics of the co-design interactions.

4 Discussions and Conclusions

Through the two studies, we observed quite different interactions between the designers and the users. Ironically, professional designers seem to perform worse than design students in co-design sessions. In the co-design workshop, professional designers played a dominant role in the whole process and controlled design directions. Although they realised the importance of understanding users, their language, attitudes and behaviour did not often show user empathy, and co-designing with users did not seem to be a natural process for them. We think the following factors at least contributed to the situation: (1) the lack of user research in the designers' education; (2) the time pressure of the co-design workshop which might have prevented designers' interactions with the users; (3) the designers might have seen the users as 'clients on the opposite side'. However, these hypotheses will need to be tested in future studies.

The Fixperts project proved that design students were much more willing to listen attentively to users and showed more respect to them after learning user research methods. They were grateful for the fixpartners' time and efforts, and regarded fixpartners as helpers to assist them in finishing their homework. This gratitude had a positive impact on the process and the result of the co-design. Nevertheless, there seems to be a natural desire on the part of students to please their teachers and to focus on the teachers' needs or desires (Strickfaden and Heylighen 2009). The students' good performance may be related to the fact that they wished to get higher scores in this module.

4.1 What have we Learned from the Co-design Studies?

4.1.1 How do Users Participate in the Co-design Process?

There are different modes of user participation (Fig. 3):

- active: users are able to articulate their needs and design ideas;
- semi-active: users are aware of their intentions/expectations;
- indirect: users do not know what they need but are willing to talk to designers;
- passive: users only give feedback when they see the design solution concepts/ prototypes, towards the end of the design process.

The four modes of user participation identified from the 'fixperts project' also apply to the co-design workshop. For example, in the three groups, the tongue-tied user (Mr. Xie) was in a passive mode. The user who considered that designers' sketching was nice but not practical was in a semi-active mode. The user who confronted the designers was in an active mode.

4.1.2 What are the Designers' Role in the Co-design Process?

Co-design requires designers to play several roles well: not only solving problems, but also managing time and process and guiding/inspiring/empowering users. Based on the two studies, we summarise the designers' roles in the co-design process in Fig. 5.

Fig. 5 Designers' roles in the co-design process

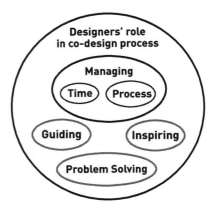

In the early stages, designers should spend time and effort on building mutual trust and good relationships with the users, especially when they are very different from the designers (e.g. age and background). During the design process, designers need to manage the process and to guide and inspire users in exploring the problem and developing solutions.

4.1.3 How can Researchers Better Support Co-design Activities?

In the co-design workshop, the researchers did not give user research advice to the professional designers as they thought experienced designers would know how to work with users. However, observation suggested this was not the case. So researchers should offer user research advice to designers in advance, as they did for the design students.

As co-design facilitators, it is also the researchers' responsibility to make all participants warm up at the early phase of the design process which will help break the ice and build trust between designers and users. In addition, as the general public in China has low awareness of design, it would be useful for the researchers to introduce the concept to users—this will make them more confident when they talk to designers and avoid unnecessary misunderstanding.

This study, although limited by its relatively small-scale and experimental nature (i.e. not real-life projects), offers a lot of insights for future studies. The authors plan to conduct further co-design workshops with professional designers to test the hypotheses and to implement the changes. The user participation modes will be further studied, tested and refined to build knowledge of co-design.

Acknowledgments The research was sponsored by the Shanghai Pujiang Scholarship Scheme. We thank all the users, designers and design students for their contribution to this study.

References

Bitner MJ, Faranda WT, Hubbert AR, Zeithaml VA (1997) Customer contributions and roles in service delivery. Int J Serv Ind Manag 8(3):193–205

Cassim J, Dong H (2007) Empowering designers and users: case studies from DBA Inclusive Design Challenge. In: Coleman R, Clarkson PJ, Dong H, Cassim J (eds) Design for inclusivity: a practical guide to accessible, innovative and user-centred design. Gower Publishing, London

Cassim J, Dong H (2013) Interdisciplinary engagement with inclusive design: the challenge workshops model. Appl Ergonom (in press): 1–5

Dong H, Yuan S (2013) Learning from co-designing. In: Proceedings of the 2nd international conference for design education researchers. Oslo, Norway

Design Council (2005) The design process. Available at: www.designcouncil.org.uk/about-design/How-designers-work/The-design-process/. Accessed on 24 Oct 2013

Design Council (2013) Co-design. Available at: www.designcouncil.org.uk/resources-and-events/designers/design-glossary/co-design/. Accessed on 24 Oct 2013

Ho DK, Ma J, Lee Y (2011) Empathy @ design research: a phenomenological study on young people experiencing participatory design for social inclusion. CoDesign 7(2):95–106

Kaasinen E, Koskela-Huotari K, Ikonen V, Niemelä M, Näkki P (2012) Three approaches to co-creating services with users. In: Proceedings of the 1st international conference on human side of service engineering. San Francisco, Canada

McDonagh D (2008) Do it until it hurts!: empathic design research. Des Prin Pract 2(3):103–110

Muller MJ, Druin A (2007) Participatory design: the third space in HCI (revised). In: Jacko J, Sears A (eds) Handbook of HCI (2nd ed). Erlbaum, Mahwah

Sander EBN, Stappers PJ (2008) Co-creation and the new landscapes of design. CoDesign (Special issue) 4(1):5–18

Sanoff H (2007) Editorial: special issue on participatory design. Des Stud 28(3):213–215

Sleeswijk VF, Stappers PJ, van der Lugt R, Sanders EBN (2005) Context-mapping: experiences from practice. CoDesign 1(2):119–149

Spinuzzi C (2005) The methodology of participatory design. Tech Commun 52(2):163–174

Strickfaden M, Heylighen A (2009) Who are they? Student voices about the other. In: Proceedings of the include 2009. Helen Hamlyn Centre RCA, London

Winograd T (1996) Bringing design to software. Addison-Wesley, Boston

Together Through Play: Facilitating Inclusive Play Through Participatory Design

R. J. Holt, A.-M. Moore and A. E. Beckett

1 Introduction

Play has an important role in the development of physical and social skills of children (Piaget 1929), and is recognised as a fundamental human right (cf Article 31 of the United Nations Convention on the Rights of the Child). Disabled children face many barriers to play, sometimes due to accessibility, but also to social barriers that arise between disabled and non-disabled children (such as ableist assumptions or sensibilities held by non-disabled children)—which are exacerbated by the difficulties of playing together.

There is a growing move towards inclusive education, encouraging the inclusion of disabled children into mainstream schools, rather than educate them separately. To be truly inclusive, such education must fully involve disabled children in all aspects of school life, including being able to play socially with their peers (cf http://inclusiveschools.org/inclusion-on-the-playground/). For play to be effective, it must be meaningful: it is not sufficient to have disabled and non-disabled children playing next to each other. They must be engaged in a way that is meaningful to both. If disabled children are denied the opportunity to engage meaningfully with others, they remain effectively excluded as well as being denied the opportunity to develop skills and exercise agency (Burke 2012).

This paper describes a participatory design project between Engineers and Sociologists at the University of Leeds that explores the aspirations of disabled and non-disabled children for playing together and the barriers that prevent this. The project takes a cooperative enquiry approach (Druin 1998), as a way of attaining a

R. J. Holt (✉) · A.-M. Moore
Institute of Design, Robotics and Optimisation, School of Mechanical Engineering,
University of Leeds, Leeds, UK
e-mail: R.J.Holt@leeds.ac.uk

A. E. Beckett
Centre for Disability Studies, School of Sociology and Social Policy,
University of Leeds, Leeds, UK

P. M. Langdon et al. (eds.), *Inclusive Designing*, DOI: 10.1007/978-3-319-05095-9_22,   245
© Springer International Publishing Switzerland 2014

richer understanding of children's views. The aim is not to develop inclusive toys per se, but to use the toys designed and prototyped as critical objects to provide insight into children's views. This chapter reviews current literature on inclusive play, describes the participatory design process used, then reviews the initial findings from this process and reflects on our experiences, including the distinction between accessibility and inclusivity in play, the role of social barriers and how these can be addressed through the design of toys and games.

The language used in this article is in keeping with the Social Model of Disability (Oliver and Barnes 2010), which views *impairment* as a property of the body and *disability* as a social relationship. According to this model, disabled people are people who have impairments, but are disabled by society. The model has not been without its critics (Allan 2010). Elsewhere Beckett (2006) has argued that the model may need to be revised to ensure that both impairment and disability are understood within a social framework. Nevertheless, it represents an important alternative to the Individual Model of Disability which views the 'problem' of disability as residing solely within the bodies of disabled individuals (Oliver 1990). It is a heuristic device that helps us to think differently about disability.

2 Inclusive Play

In recent years, campaigns by disabled people's organisations such as 'Disability Equality in Education' (http://www.diseed.org.uk) and the 'Alliance for Inclusive Education' (http://www.allfie.org.uk) have brought a shift towards inclusive education in the UK. As a result of the Special Educational Needs and Disability Act (2001) there is now greater representation of disabled children in mainstream schools in the UK. Inclusion is, however, about more than simply 'integrating' disabled children into the physical environment of the mainstream classroom; it is also about ensuring that schools and classrooms become 'inclusive' environments in their values and ethos (Barton and Armstrong 2007).

Play is an important aspect of this integration or inclusion—where children are not able to play *meaningfully* together, they cannot be said to be truly included. A range of efforts have been made in terms of making toys and environments accessible to children with impairments, though these have generally focussed upon outdoor play areas (e.g. http://www.inclusiveplay.com). Endicott et al. (2010) have adapted the principles of universal design to develop guidelines for the design of inclusive playthings and environments, and undertaken some comparisons of the differences in the way that disabled and non-disabled children play (Endicott et al. 2009). However, the emphasis remains on making play accessible to children—which is not the same as ensuring that children are included in play. Instead, this implies a social aspect, willingness for disabled and non-disabled children to play together in a way that is meaningful to both. The challenges experienced by disabled children are not merely those of accessibility: their lack of power makes them more vulnerable to the views of society, which encourage low self esteem and 'internalised oppression' (Reeve 2004). This project places an emphasis on the social and emotional aspects which make play opportunities *meaningful* to children (Golinkoff et al. 2006). For this project, meaningful play is play that allows children to establish friendships, have positive interactions with peers and others; empowers disabled children, challenging processes that lead to internalised oppression (ableism); challenges perceptions about impairment/disability and any ableist assumptions held by non-disabled children.

There is some evidence that increased contact between disabled and non-disabled children has a positive effect on attitudes between them (Maras and Brown 1996); but also evidence that increased interaction with disabled peers can increase negative attitudes (Hodkinson 2007). Despite general agreement within Disability Studies that negative attitudes towards disabled people—often termed 'disabling attitudes'—are a significant problem, little *empirical* research, informed by Disability Studies perspectives, has been conducted to explore these attitudes, especially as they are held/articulated by children. There is much *theorising* about disabling attitudes. Authors within Disability Studies have mostly rejected the traditional 'psychological' approaches which tend to view attitudes as formed by individuals in isolation (Howarth 2006), rather considering ideas about disabled people to take shape 'in interaction, in dialogue and in practice with others' and to be 'anchored in traditions and ideologies' (Howarth 2006).

Play is an example of just this sort of 'interaction, dialogue and practice' and we should be concerned about the way this is taking place within lay situations, lest it reinforce disabling attitudes. Our study seeks to find new ways to address this. In this project, participatory design, prototyping and testing are the vehicles for exploring the question: can play provide an opportunity for children (disabled and non-disabled) to form positive ideas about disabled people?

3 Participatory Design Process

The challenges of doing research and design with children are well-documented (Markopoulos et al. 2008), not least because of the traditional power relationship between adults and children. Druin (1998) proposed the concept of cooperative inquiry: participatory design with children not as a way of generating great designs, but as a way of better understanding their views. Researchers on the project are working with groups of disabled and non-disabled children to develop and evaluate designs. The aim is not to design inclusive toys and games per se, but to identify children's aspirations for playing together, the barriers that prevent this, and how they might be overcome.

The project has adopted a series of iterative cycles:

1. Initial group interviews with children about the experience of play;
2. Working with children to develop initial concepts;
3. Building lo-fidelity prototypes to illustrate game concepts for evaluation;
4. Revising the concepts based on feedback, and developing hi-fidelity prototypes that children can use to play;
5. Selection and refinement of two most preferred concepts, for final evaluation with children.

In this way, we not only explore their ideas, but also test them out. It is important to understand the reasoning behind comments and preferences, and to explore them more deeply. Design and evaluation sessions were recorded for detailed coding and analysis, in exactly the same way as the interviews, with the prototypes becoming a probe for prompting discussion.

Twenty-two children at four schools are participating in the project in the seven to ten age range. The six disabled children recruited to take part in the project have physical impairments relating to cerebral palsy (for four of the children), hearing impairment (deafness) and dyspraxia. Children were recruited at the discretion of the school in small friendship groups of three to six in size, such that at least one child in each group had an impairment (as recognised in their statement of Special Educational Needs), and at least one did not. Informed parental consent was required for all children who participated, as well as assent from the participating children themselves. Clearly, there are a huge range of potential impairments, and it would be impossible to systematically cover them all. The scope of this project was limited to physical impairments, and the sample is in many ways a convenience sample taken mainly from schools who have worked with the University previously. The aim was to conduct an exploratory study to identify children's views, rather than a systematic and representative study. At the time of writing, data gathering for the first four stages has been completed and it is the outcomes of this that are reported.

4 Results

This section reviews the work completed at the time of writing, and the initial findings of the project. It is worth noting that a detailed analysis of the interviews and feedback sessions which will offer a richer interpretation of children's attitudes towards play and inclusivity is still underway, and will only be completed once testing with the finalised prototypes is complete. The results presented here describe the design outcomes, and our main observations from testing.

4.1 Initial Interviews

Semi-structured group interviews were conducted with each friendship group, to identify their preferences for play, and any experiences they had of exclusion. In addition, the children were asked to brainstorm ideas for games which disabled and non-disabled children could play together. There was no expectation that the children's designs would be potential solutions (though if they were, so much the better)—rather, they provide a further source of data about their play preferences, and their attitudes towards impairment and disability.

The disabled children were able to give several examples of being excluded from play as a result of impairments. Being physically slower and the preconceptions of other children were both identified as issues, and for the child with impaired hearing, difficulty in understanding the game being played and having to ask for help or clarification were major barriers. Among the children more generally, age was identified as a major factor, with older children excluding younger children, particularly siblings, and more popular children excluding those they perceived as less popular. It was also noted that such exclusion could be extremely mean, with some children noting the insults used to drive away excluded children, and some explicitly characterising it as bullying, and noting how upsetting this exclusion could be for the child involved.

Unfairness and excessive dominance were particular problems. Children disliked it when one or more children 'took over' and 'spoil(ed) a game', leaving the others feeling left out. Dishonesty was also seen as something that spoiled play and children indicated that they tended to avoid playing with children who they thought would not play fair. All the children reported that they strongly disliked falling out with friends over games, and that this was the thing they most disliked about play. It was noted, however, that sometimes children just had different aspirations for play, and that they might decide not to play because the group as a whole was not playing the game they wanted.

Lack of confidence was also noted as a potential cause of exclusion, with children feeling unable to approach groups to join in, or not wanting to participate in a game that they did not feel sufficiently good at. Children also tended to want to play with children of similar ability. Team sizes were also an issue, with children citing a range of games from Rugby to Connect 4 as problematic because only so many children could play, meaning that any other children were automatically left out. Being perceived as different was also seen as problematic, and it was noted that children who had to rely a lot on a parent, often felt left out when playing in public spaces, where other children could gravitate towards each other and play together.

Children's designs tended to be derivations of existing games, particularly videogames such as SingStar^{TM}, Minecraft and Call of Duty (despite this last being rated as unsuitable for children). One group designed a piece of outdoor play equipment called the *Fort of Doom*, and devised an elaborate series of games around it, though they did indicate that they would also be happy for it to be a videogame. Customisability and collectability also featured prominently, with references to toys such as Moshi Monsters^{TM}. In terms of accessibility, the non-disabled children tended to opt for a system whereby disabled children would be given more turns, or a golf-style handicap, because they 'wouldn't be as good'. Some of the disabled children expressed a desire for games that could be quite physically challenging—for example, one child with impaired arm function due to cerebral palsy designed a game based around solving puzzles which required quite fine dexterity, because this was the sort of game he wanted to play. Notably, however, he stipulated that the game should be playable one-handed, allowing him to use his unimpaired arm, and removing the key barrier to his participation.

4.2 Concepts and Low Fidelity Prototyping

Based on the children's initial ideas, five game concepts were developed in conjunction with a team of undergraduate Product Design students at the University of Leeds. These drew upon the concepts developed by the children for inspiration, though considerable adaptation was required to ensure they conformed as far as possible to Endicott et al. (2009) adaptation of the Universal Design Guidelines, and some ideas were merged. The aim was to design specifically for the tastes and capabilities of the 22 children participating, and whose aspirations we were exploring, rather than trying to create truly universal toys. *3D Stack* engaged children in the task of building a tower from shaped blocks, with the aim of building the highest tower possible; *Jump On* was a videogame controlled by pressing buttons on a mat, such that children could steer a hovercraft in the game by moving around the mat to change the balance of the craft; *Battle Balls* was a modern spin on conkers, using larger 'monster' heads with the aim of striking the opponent's target and detaching it from their string; *Escape the Castle* was an educational board game, in which the children would move around a board and carry out asks related to different subjects (Maths, Art, etc.) in order to escape from the fictional castle as a team; and *Puzzled* presented a two player memory game, in which one player would press out a sequence of buttons, causing lights to flash on the other player's side of the board, and the other player would have to reproduce the sequence within a time limit. Low fidelity models were produced, as shown in Fig. 1, and taken into schools for discussion. The prototypes were not functional, although it was possible to simulate play with them: their purpose was to help communicate the ideas to the children for discussion before they were developed further, rather than practical testing. In this they were quite effective, though children struggled to imagine some of the functions and this was reflected in their feedback. The children were generally positive about the games (although it should be noted that children show a bias towards positive feedback (cf Markopoulos et al. 2008)).

Children were keen to have a team-based approach in *3D Stack*, with two teams competing with their own pieces to build the highest tower, each within its own footprint. The teams should also be multiplayer: 'games are more fun when you

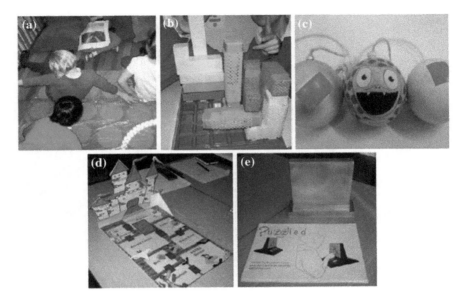

Fig. 1 Low fidelity prototypes: **a** Jump on; **b** 3D Stack; **c** Battle Balls; **d** Escape the Castle; **e** Puzzled

can include more people'. A time limit was also a popular suggestion, though there was a lot of disagreement about how long it should be, suggestions ranging from 5 min to half an hour. Alternative names were also proposed: 'Stackamo' or 'Stackcraft' (in honour of Minecraft) were popular choices.

Jump On was popular for the fact that it involved a videogame, and the idea of collaborating to steer a vehicle was popular—but the mode of interaction did not work well. Children did not like having to sit in such close proximity, and some of the disabled children found sitting or reaching to the side particularly difficult, noting that it would be better if the layout of controls was more flexible.

The children had a lot of fun with *Battle Balls*, but found that the strings tended to get tangled, and the children with arm impairments found it difficult to use the low fidelity prototypes, because they were bi-manual (one hand to hold the string, and one hand to aim and fire). They suggested that it would be good to have a more rigid wire, to make them easier to use. It was also suggested that the characters should look more monstrous: the boys in particular were concerned that the Battle Balls looked too cute and colourful.

Escape the Castle was less popular because of its educational aspect, and the children felt that the castle did not look spooky enough and that they were moving around the castle rather than escaping from it. They enjoyed the mix of activities, but this depended on their abilities—those who were less good at maths, for example, disliked having to do maths questions.

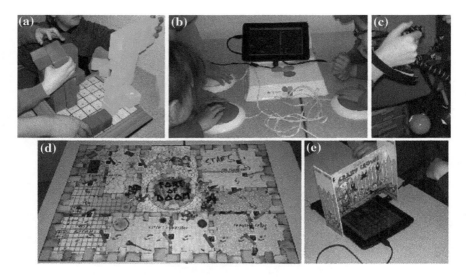

Fig. 2 Functional prototypes of **a** Stackamo; **b** Button Bash; **c** Battle Balls; **d** Escape the Fort of Doom!; and **e** Puzzled: Crazy Crows

Puzzled was also one of the less popular concepts, with some children being keen (particularly those who had suggested this idea in the first place), but others were concerned that it was quite boring, or that the child setting the pattern might deliberately make it too difficult for the other child.

Children were particularly interested in how these games were based on their ideas, and some asked if they would be able to make their own prototypes. On the whole, though, they were very keen on being consulted about design decisions: 'Can I say something? In my head I feel like I'm in this grown up meeting, deciding about complicated engineering'.

4.3 Functional Prototypes

Based on these comments, the games were refined and functional prototypes were built to allow the children to try them out, as shown in Fig. 2. *3D Stack* was renamed *Stackamo* at the children's suggestion, and the static board was replaced with a set of LEDs that could light up in blue or green, giving separate footprints for two teams to build their own towers with the appropriate coloured blocks. *Battle Balls* were refined with more monstrous faces, a strap in place of a string (to avoid tangling and encourage a straighter trajectory) and a grip that allowed attachment to a table or forearm, rather than just being held in the hand. *Escape the Castle* was renamed *Escape the Fort of Doom!* at the children's suggestion. The board was made more 'scary' and it was made clearer that players would be heading towards the exit. *Puzzled* was given a 'crazy crows' theme, based around crows raiding a cabbage patch, in order to make it more visually interesting, and was implemented as a computer game using National Instruments' LabVIEW™ on a tablet computer, with a physical dividing screen to separate the two halves.

Jump On was the only concept to be significantly altered due to feedback on the low fidelity prototype. The mat concept was abandoned in favour of the use of tactile switches (similar to the Tash Buddy Button), the game again being implemented using National Instruments' LabVIEW™. The game moved away from the concept of steering, closer to Wackamole, where each player had to press their button when the relevant colour of alien popped up. Each player was given a score and the team accumulated a score as a whole.

The children were given the opportunity to play with the prototypes, and were given a vote on which they would most like to see developed. *Button Bash* and *Battle Balls* came out as the two most popular, followed by *Stackamo* and then *Escape the Fort of Doom!*, with *Puzzled* being the least popular. The *Escape the Fort of Doom!* theme was popular, but some children did request that it should be a videogame, rather than a board game.

The physical accessibility of the prototypes was good, with the exception of the mechanism for reattaching the *Battle Balls* after they had been triggered. This required significant manual dexterity and proved problematic for the children who were impaired in this area.

Most significantly, the alterations made in transforming *Jump On* into *Button Bash* may have improved its physical accessibility, but the alterations were not all positive. The inclusion of individual scores meant that there was immediately an element of competition between the children, with some complaints that the game was 'unfair' because one player got more aliens presented to them than another (even though this was factually incorrect!). Some children felt frustrated if they were not able to get what they perceived as a good enough score. Most significantly, the limited number of buttons meant that not every child could play.

One observation was that where only a limited number of players could participate in a game, social pecking orders asserted themselves, and the disabled children tended to be at the back of the queue, meaning that while they were physically able to play the games, they were still socially excluded until the research team intervened. Where there was space for everyone to play (the team-based approaches to *Stackamo* and *Escape the Fort of Doom!* for example), this did not happen, although in some cases there were unpleasant arguments of the form 'I'm not being a team with him/her!', vividly demonstrating the sort of unpleasant behaviour children had mentioned in the interviews.

Interestingly, *Battle Balls*—which was both two player and competitive—did not seem to experience this problem. Our interpretation is that this was because games were quick, and there were enough resources for everyone to have a Battle Ball, and so take turns to compete. It was also quite an effective spectator sport, with children cheering each other on and getting quite involved even when they were not playing. By comparison, computer-based games such as *Puzzled* and *Button Bash* were generally watched in silence.

5 Conclusions and Further Work

This chapter provides an overview of the work carried out on this project to date and our main observations. The main lessons are the significance of social barriers as well as physical barriers to inclusion: i.e. a game may be accessible, yet the behaviour of the players determines whether or not it is inclusive. It seems that these social barriers may be removed (or could be ameliorated) through better design—such as allowing variability in the number of players and encouraging collaboration rather than competition. Of course, this then raises the question of whether avoiding competition is a healthy approach, or whether there may be ways of encouraging more 'constructive' competition.

The final step of prototyping has yet to be undertaken: both *Battle Balls* and *Button Bash* will be further refined based on feedback from the functional prototypes and taken back to the children for further evaluation. In *Battle Balls*, the trigger mechanism will be refined to ensure easier reattachment. *Button Bash* will again be extensively redesigned, to accommodate a variable number of players, to emphasise a more cooperative playing style, without individual scores and to adopt the *Escape the Fort of Doom!* theme. This will allow us to assess whether these changes can mitigate the social barriers that arose through the design in the functional prototype testing.

It is also worth noting that the detailed coding and analysis of the interviews and prototype testing have yet to be conducted. These have provided a very rich source of data now being complemented by data on inclusive play from teachers and parents. Through the analysis of these we will be better able to grasp whether inclusive play can encourage the development of positive attitudes between disabled and non-disabled children.

Acknowledgments This project was funded through a Leverhulme Trust Research Grant. The authors wish to thank the participating children; the parents and teachers who helped to arrange this; the students who aided the design and construction of prototypes: Matt Hodgkinson, Liam Kilkenny, Rik Loong, Billy Pagett and Alex Styles; and also Abbas Ismail who provided technical support in the manufacture of the prototypes.

References

Allan J (2010) The sociology of disability and the struggle for inclusive education. Br J Sociol Edu 31(5):603–619
Beckett AE (2006) Citizenship and vulnerability: disability and issues of social and political engagement. Palgrave Macmillan, Basingstoke
Barton L, Armstrong F (2007) Policy, experience and change: cross-cultural reflections on inclusive education. Springer, Dordrecht
Burke J (2012) Some kids climb up; some kids climb down: culturally constructed play-worlds of children with impairments. Disabil Soc 27(7):965–981
Druin A (1998) The design of children's technology. Morgan Kauffman, San Francisco

Endicott S, Mullick A, Kar G, Topping M (2010) Development of the inclusive indoor play design guidelines. In: Proceedings of the 54th annual meeting of the human factors and ergonomics society. San Francisco, Canada

Endicott S, Kar G, Mullick A (2009) Inclusive indoor play: children at play. In: Proceedings of the 53rd annual meeting of the human factors and ergonomics society. San Antonio, Texas

Golinkoff RM, Hirsh-Pasek K, Singer D (2006) Play=learning: a challenge for parents and educators. In: Singer D, Golinkoff RM, Hirsh-Pasek K (eds) Play=learning: how play motivates and enhances children's cognitive and social-emotional growth. Oxford University Press, New York

Hodkinson A (2007) Inclusive education and the cultural representation of disability and disabled people: recipe for disaster or catalyst of change? An examination of non-disabled primary school children's attitudes to children with disabilities. Res Educ 77(1):56–76

Howarth C (2006) How social representations of attitudes have informed attitude theories: the consensual and the reified Theory and psychology, 16(5):691–714

Markopoulos P, Read J, MacFarlane S, Hoysniemi J (2008) Evaluating children's interactive technologies. Morgan Kaufman, San Francisco

Maras P, Brown R (1996) Effects of contact on children's attitudes toward disability: a longitudinal study. J Appl Soc Pyschol 26(23):2113–2134

Oliver M (1990) The politics of disablement. Macmillan Education, London

Oliver M, Barnes C (2010) Disability studies, disabled people and the struggle for inclusion. Br J Sociol Educ 31(5):547–560

Piaget J (1929) The child's conception of the world. In: Tomlinson J, Tomlinson A (trans). Routledge, London

Reeve D (2004) Psycho-emotional dimensions of disability and the social model. In: Barnes C, Mercer G (eds) Implementing the social model of disability: theory and research. The Disability Press, Leeds

Part VI
Legislation, Standards and Policy

Inclusive Design of a Web-Based Resource to Assist Public Librarians with Providing Government Information to Library Patrons in the United States

B. Wentz and J. Lazar

1 Introduction

Citizens often go to their public libraries for assistance with accessing government information and services, and this results in public librarians being under high demand to answer their questions. Given the rapid turnaround times expected by library patrons, as well as the domain knowledge needed by librarians in the area of government, a web-based resource was created to assist librarians who are helping citizens find and understand information about government. Due to the diversity of public libraries (and librarians) in the United States, this website needed to be highly usable for all public librarians, as well as accessible to librarians with disabilities. This paper documents the inclusive design and evaluation of the 'LibEGov' project involving citizens, librarians, developers and inclusive design researchers to understand the specific challenges of developing a usable and accessible web-based resource.

2 Related Literature

For many citizens in the US, the UK and other countries, their primary method for interacting with government services is through websites, often known as 'e-government' (Bertot et al. 2009; European Commission 2011). Common

B. Wentz (✉)
Department of Management Information Systems, Shippensburg University,
Shippensburg, USA
e-mail: bjwentz@gmail.com

J. Lazar
Radcliffe Institute for Advanced Study, Harvard University, Cambridge, USA

J. Lazar
Department of Computer and Information Sciences, Towson University, Towson, USA

P. M. Langdon et al. (eds.), *Inclusive Designing*, DOI: 10.1007/978-3-319-05095-9_23,   259
© Springer International Publishing Switzerland 2014

questions or tasks requiring interaction with e-government websites include immigration or citizenship information and tax information. A challenge to e-government usage is that 19 % of US adults are not Internet users (Pew Research Center 2012), and even individuals with Internet access at home face challenges with locating information, understanding government structure and language barriers (Jaeger and Thompson 2004). Over time, government agencies have been providing less in-person or phone service, instead directing people to their websites. It has become commonplace for US government agencies to direct citizens with questions about government agency websites to instead visit their local public libraries (Fisher et al. 2010). From a public library perspective, this creates complex challenges, and there is a need to assist librarians at public libraries, who themselves are assisting patrons with e-government services. This paper describes the process of utilising inclusive design methods to develop a website to assist librarians with providing e-government services to the public.

3 Research Methodology

The development of the LibEGov website utilised a fully inclusive and user-centred design process. A number of preliminary data gathering activities took place in 2011, including interviewing representatives of five US federal agencies and site visits to seven public libraries (in Connecticut, Florida, Georgia, Maryland and Texas). At each library site visit, state library staff, government officials and members of the community were all interviewed (the libraries were a combination of rural, urban and suburban libraries, with different socioeconomic levels). In addition, the results of the 2011–2012 Public Library Funding and Technology Access Survey (PLFTAS) were also analysed (Bertot et al. 2012). All of those data collection efforts are documented in (Jaeger et al. 2013) and were utilised to help build a very preliminary prototype of the LibEGov website, in March 2012. This paper focuses on the user testing and accessibility evaluations that took place throughout 2012. To help effectively meet the needs of multiple user populations, the LibEGov website underwent expert reviews, accessibility evaluations and extensive user testing with diverse users at a variety of library locations. The initial data collection efforts identified key design elements to be included (such as the ability to create an account and provide a forum for discussion), and content areas (at first, focusing on the areas of immigration and taxation). The entire process was iterative, with feedback from librarians provided to the developers on a weekly and sometimes bi-weekly basis. Changes were often made within hours of suggestions being made and were evaluated by additional librarians only days later. This was to ensure that the perspective and user experience of the public librarians—the intended users of the site—was central to the completed LibEGov resource. The overall process was as follows:

1. March 2012: Expert reviews and accessibility inspections.
2. April and May 2012: First round of user testing with 35 library staff.
3. June 2012: Major revisions and content added to the LibEGov site.
4. July and August 2012: Second round of user testing with 35 library staff.

3.1 Expert Reviews and Accessibility Inspections

The evaluation of the site began with expert site reviews by the researchers providing early feedback to developers. Expert reviews are a standard part of user-centred design methods, and are typically conducted prior to user testing, because expert reviews can find larger, more obvious interface flaws, allowing the developers to fix those flaws before user testing begins (Lazar 2006). User testing can then identify less obvious flaws more closely related to the tasks and user knowledge, which experts might not be able to identify. In addition to expert reviews for general usability, public library websites and websites funded by government agencies in the US must be accessible for users with disabilities (Lazar et al. 2011). The accessibility inspections of the LibEGov site, occurring at the same time as the expert reviews, involved using a structured, 3-phase approach to evaluate its accessibility by (1) listening to a web page with a screen reader while visually viewing the page, (2) using typical non-visual navigation approaches, and (3) inspecting code to ensure compliance with each paragraph of the US Section 508 interface regulations. This widely accepted approach for accessibility evaluations ensured compliance with the law, addressed different types of disabilities (e.g. the keyboard access required by screen reader users is also what users with motor impairments need), and has been used in previous evaluations (Wentz et al. 2012), including evaluations of library websites (Lazar et al. 2012). The homepage, primary menu pages, and one level below primary items (where applicable) in the information architecture were evaluated for accessibility throughout the project, as well as common activities such as registering for an account and adding links to the 'Bookmarks' page.

3.2 User Testing

Based on the content available on the LibEGov portal at the time of the user testing (starting in April 2012), the researchers decided to focus the user tasks on the following four areas: user account registration and functionality, information seeking, bookshelf functionality and blog/forum usage. Prior to the first round of user testing, a pilot study of an initial task list (19 tasks) took place with a librarian who was not taking part in any of the user testing at the seven library systems in the state of Maryland, and some tasks were clarified as a result. Since the user testing took part over a two month period, changes were immediately made and evaluated by subsequent users, in an interactive design process. For the second phase of user testing (July–August 2012), none of the functionality or usability on the user account registration, bookshelf or blog/form had changed since the end of May 2012. The only substantial changes since the first round of user testing were in the content areas of taxation and immigration, where much more content had been added. Therefore, the tasks during phase two focused solely on information seeking for content in the taxation, immigration and training areas. A new task list (17 tasks) was tested with a pilot participant who was not taking part in any of the user testing at the seven library systems.

The user testing was conducted on-site at the individual libraries or library offices using library computers to accurately emulate the typical library work environment (except without a line of patrons waiting to ask questions, which would, in reality, add an additional time pressure). Before each user evaluation began, the researchers emphasised that it was the interface which was being evaluated and not the user, and users were encouraged to provide feedback at any point. The process involved an informed consent agreement, a survey about experience with providing e-government services, timed and observed tasks on the LibEGov interface, and answering some wrap-up questions at the end of the evaluation. Throughout the user testing, users were asked to only look for the answers to tasks within the LibEGov.org website, even if they thought that the information was available on an outside website, because the user testing was focusing on LibEGov.org and not another site.

For each of the two rounds of user testing, 35 library staff used and evaluated the LibEGov web portal, from a mix of urban, suburban and rural libraries (five users for each of the seven library systems involved). In addition, two of the users who took part in each phase of the user testing were blind and used the JAWS screen reader. The approach used in this evaluation met or exceeded the generally accepted number of users (Turner et al. 2006). The average age of the users in phase one was 46.1 years and 44.6 years during phase two (range 26–68; one user did not wish to reveal their age). The average number of years at their current library system in phase one was 10.6 years and 9.9 years during phase two (range 1–42 years). There were 29 females and six males during phase one (26 females and nine males during phase two). In phase two, 27 of the 35 users had also taken part in the user testing in April–May. While it was not always possible to get the same users in the two phases of user testing, the demographics of participants in both rounds were very similar.

4 Results

4.1 Initial Expert Reviews and Accessibility Inspections

The researchers provided the development team with multiple sets of design suggestions from the initial expert reviews and accessibility inspections, before the user testing. Examples of the feedback given included the following:

- There were areas of poor colour combinations affecting readability.
- It was not clear how the blog and forum were different.
- There was a lot of wasted space on the header area of each page.
- There needed to be a site search feature located near the top of each web page (preferably with an auto-suggest feature).
- Users needed to be aware that the login names were case-sensitive.

Some of the feedback specific to the accessibility inspections included:

- A consistent 'skip navigation' link needed to be at the top of every page.
- For the navigation menu, while drop-down menus can theoretically be accessed without a mouse, there needed to be feedback when a link with a drop-down menu is selected, so users know that sub-menus exist.
- On all the pages, the tab order jumped first to the breadcrumbs navigation before the main links, and it should not have, since that was not the order in which it appeared visually on the page.
- Confirmation for items added to the bookshelf was not accessible for screen reader users (it displayed for a few seconds, then disappeared).
- PDFs should include a link to a viewer. Also, if a PDF file is not accessible, an alternate (such as HTML for the content) should be included.

4.2 User Testing Results

4.2.1 Phase One Task Performance

Caution is needed to be paid in interpreting the results of task performance, since the users were evaluating different versions, for instance, with minor changes made after the first five users, and major changes made after the first 20 users.

Generally, the users were able to achieve a high level of success on most of the tasks. The only tasks that had a success rate much below 90 % were Tasks 6 and 7, which were tasks related to searching for immigration information. Task 6 had a completion rate of 71 % (25/35) since users were confused when they looked for the information on the Immigration page. Users skipped right over the tab links because they only exist on this one page within the site and did not notice a new navigation area that only exists on this one section of the site. Note that Tasks 2

Table 1 Task completion results for phase one

Task	Task summary	Performance (number of users, percentage)	
1	Register for an account on LibEGov	35/35	100 %
2	Log into LibEGov, then create a permanent password	32/33	97 %
3	Log out of LibEGov	33/33	100 %
4	What is the highest income level to use free IRS online filing?	31/35	89 %
5	How long must taxpayer wait to check refund status on IRS site?	32/35	91 %
6	What form does immigrant use to apply for residency?	25/35	71 %
7	Name a site that LibEGov suggest for immigration legal services	27/35	77 %
8	What are three types of basic literacy suggested for patrons?	33/35	94 %
9	Number of success factors for library partnerships on LibEGov?	31/35	89 %
10	Log back into your account	35/35	100 %
11	Add information to your account profile	31/35	89 %
12	Use a contact form to send the LibEGov team a message	35/35	100 %
13	On immigration, add Legal 'Ready Reference' to 'Bookshelf'	33/35	94 %
14	On taxation, add E-filing 'Ready Reference' to 'Bookshelf'	31/35	89 %
15	Go to 'Bookshelf' and locate the E-filing link	35/35	100 %
16	Remove Legal 'Ready Reference' link from 'Bookshelf'	35/35	100 %
18	Post a new message in the Taxation Forum	35/35	100 %
19	Log out of your account	34/35	97 %

Note Task 17 asked users about their perceptions about the difference between the 'blog' and the 'forum', so there was no task performance available

and 3 had only 33 users due to email provider problems at the time of the testing. Results of the phase one task performance are summarised in Table 1.

4.2.2 Phase One Design Suggestions

After each week of user testing (and sometimes, mid-week), a summary of potential changes was compiled based on problems identified, so that the development team could make immediate changes, allowing the next user(s) to evaluate the changes and confirm if these were successful. This continuously iterative approach, where problematic patterns appear and an interface is immediately modified, is considered appropriate in industry, and is what helps distinguish user testing from a more controlled experimental design (Romano-Bergstrom et al. 2011). The feedback based on interaction patterns that appeared or became clearer, included the following examples:

1. Users looked at the 'Add to Bookshelf' link and assumed that that link applied to the same item as the 'read more' link (Fig. 1).
2. After users changed their password, they were presented with the same password screen, so they thought that they either needed to re-login, or change the password again.

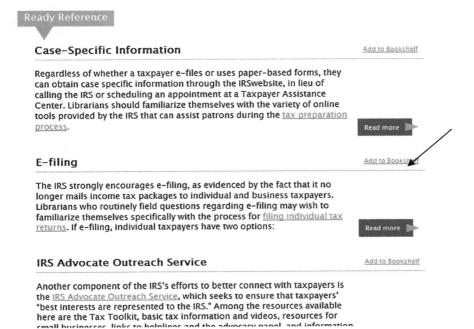

Fig. 1 Confusing location of 'Add to Bookshelf' links. Its location above the visual line makes users question which item is being added to the bookshelf

3. Users noted that it was unclear which links were internal to the LibEGov website and which links were external links to other websites.
4. Users noted that the search engine was not as flexible as Google (e.g. close matches, complete results and 'did you mean?').
5. The site did not clearly state when content was last updated. Because government content changes often, users would likely go directly to the government site instead of LibEGov if content was out of date.
6. The users disliked the wordiness and text-heaviness of the site. Users preferred text that was easier to scan.

4.2.3 Phase Two Task Performance

On 30 and 31 July 30 2012, seven users took part in user testing with the enhanced version of the site, containing all the content added in June and July (six of these users had also taken part in the April–May testing). The results of this evaluation highlighted significant problems and user complaints with the site. The new content seemed to be presented in an inconsistent and confusing manner. For instance, information about classes was found under 'classes', 'programming',

Table 2 Task completion results for phase two (final 28 users)

Task	Task summary	Performance (number of users, percentage)	
1	Class modules related to taxation on LibEGov?	20/28	71 %
2	What age group is the 'Money Math' training module for?	26/28	93 %
3	Eligibility requirements for VITA and TCE?	28/28	100 %
4	Form for immigrant adjusting status to permanent residency?	21/28	75 %
5	Organisation for non-profit free/low-cost legal services?	23/28	82 %
6	Six objectives for 'Immigration 101' course plan?	26/28	93 %
7	Six libraries with best practise for classes on immigration?	27/28	96 %
8	How is practise test different from real civics test?	25/28	89 %
9	Website in Spanish on avoiding immigration scams?	N/A	N/A
10	Section of 'Code of Federal Regulations' on immigration?	26/28	93 %
11	Policy instructions for printing boarding passes?	N/A	N/A
12	Advice when patrons ask librarians to click on e-signatures?	23/28	82 %
13	How to 'set the tone' when assisting with e-government?	23/28	82 %
14	Baltimore public library effort on 'food deserts'?	27/28	96 %
15	Website for government information on loans and debt?	24/28	86 %
16	Disclaimer language on 'legal advice versus information'?	27/28	96 %
17	When will Treasury require Social Security via direct deposit?	6/28	21 %

'training', 'best practises' and other keywords. Legal information was available under multiple headings in different places. In addition, the website still had some of the same problems that were noted in the earlier user testing: long pages of text which were not broken up or easier to navigate.

The two blind users noticed problems that continued to remain on the site: the 'click here' and 'read more' links remained on several pages, and did not clearly identify links for screen reader users. Also, since Word and PDF documents were provided on the site, links to document viewers should also have been provided. After receiving all of this urgent information, the development team asked for a delay in the user testing, which was originally supposed to continue immediately. The overall task performance for the first seven users, at the end of July, was 73.1 % (87 tasks/119 attempted), which is not considered good for user testing.

It is important to note that, during modifications to the website, the content of two tasks (9 and 11) were removed from the site, so users who evaluated the site in August 2012 were not able to attempt those tasks, and those tasks were removed from the task list. Because the site changed drastically between the first seven users who tested the site in late July and the other 28 users who tested in mid-to-late August, the researchers only included the performance results for the 28 users who took part after the major site modifications. Results of the phase two task performance are summarised in Table 2.

The average task performance for the 28 users who performed user testing after the site modifications was 83.8 % (352/420). This was an improvement over the 73.1 % average task performance from the first seven users in July 2012. It is also important to note that the only task that had below a 70 % task success rate was

Task 17, which was to find a posting in the forum that was not possible to find with the search feature (it was unable to be corrected in time for the final 28 users). If that one task is not included in the data, the task performance for the 28 users who took part in user testing in August was 88.3 % (346/392), more than a 15 % increase in task success from the users who took part in user testing in July.

4.2.4 Phase Two Design Suggestions

As was the standard practise during the previous April–May 2012 user testing, the evaluators regularly sent design suggestions throughout the user testing process, based on their observations and often on a weekly basis. The development team was sometimes able to make modifications, and at other times was not able to make complex modifications in time for subsequent users. The following are examples of design suggestions following the user testing:

- A number of users had trouble reading the breadcrumb navigation because of the colour and size of the font.
- Users continued to ask for fewer long blocks of text, and more headings, bulleted lists, bold and italicised text, etc. to help bring important content to the attention of librarians quickly.
- Users also seem to always look for common forms under links called 'Forms' in the immigration section (most forms were not under that link).
- It should be noted in a prominent way that this is not an official government website.

5 Discussion

In many ways, the user testing was very successful in identifying problems during the iterative development of the interface. In late July 2012, the first few users testing the website identified that there were major usability problems. The development team correctly decided that major changes were needed, and the user testing was delayed while the content on the website was restructured. While this makes the results of the user testing more complex to interpret, the important goal of all user testing is to identify interface flaws that can be improved upon, not to have strict analytics which can be easily compared. Again, that is what separates user testing from true experimental design.

Overall task performance did improve after the site was restructured in early August 2012. However, the limitations of the search engine, which kept most users from finding information posted in the site forum, were still present, and users still found that the content is too wordy with only limited features existing to assist the users in skimming and searching text. Users also complained about inconsistencies on the site and too many unexplained acronyms. The headings and labels were still considered to be confusing and unclear, and some of the accessibility problems remained. Since the conclusion of the user testing, the developers have continued to work on modifications to the site to make it more accessible and usable.

The inclusive design process included a high level of stakeholder involvement—government officials, librarians and community members took part in data collection. Librarians at a diverse set of public libraries performed 70 user tests on the LibEGov site at various stages. The user testing, as well as the expert reviews and accessibility inspections, helped to improve the site and ensure ease of use for the targeted users—public librarians.

References

Bertot J, Jaeger P, Shuler J, Simmons S, Grimes J (2009) Reconciling government documents and e-government: government information in policy, librarianship, and education. Gov Inf Q 26:433–436

Bertot J, McDermott A, Lincoln R, Rea, B, Peterson K (2012) Public library funding and technology access survey: survey findings and results. Available at: http://plinternetsurvey. org/sites/default/files/publications/2012_plftas.pdf (Accessed on 11 Nov 2013)

European Commission (2011) e-Government in the United Kingdom. Available at: http://www. epractice.eu/files/eGovernmentUnitedKingdom.pdf (Accessed on 11 Nov 2013)

Fisher K, Becker S, Crandall M (2010) E-government service use and impact through public libraries: preliminary findings from a national study of public access computing in public libraries. In: Proceedings of the 43rd Hawaii international conference on system sciences

Jaeger P, Thompson K (2004) Social information behavior and the democratic process: information poverty, normative behavior, and electronic government in the United States. Libr Inf Sci Res 26(1):94–107

Jaeger PT, Gorham U, Bertot JC, Taylor NG, Larson E et al. (2013) Connecting government, libraries, and communities: Information behavior theory and information intermediaries in the design of the LibEGov tool. In: Proceedings of the iConference 2013, Fort Worth, TX, US

Lazar J (2006) Web usability: a user-centered design approach. Addison-Wesley, Boston

Lazar J, Jaeger P, Bertot J (2011) Persons with disabilities and physical and virtual public library settings. In: Bertot J, Jaeger P, McClure C (eds) Public libraries and the Internet: roles, perspectives, and implications. Libraries Unlimited, Santa Barbara

Lazar J, Wentz B, Akeley C, Almulhim M, Barmoy S et al (2012) Equal access to information? Evaluating the accessibility of public library websites in the State of Maryland. In: Langdon PM, Clarkson PJ, Robinson P, Lazar J, Heylighen A (eds) Designing inclusive systems. Springer, London

Pew Research Center (2012) Demographics of Internet users. Available at: http://pewinternet.org/ Static-Pages/Trend-Data-(Adults)/Whos-Online.aspx (Accessed on 11 Nov 2013)

Romano-Bergstrom J, Olmstead-Hawala E, Chen J, Murphy E (2011) Conducting iterative usability testing on a web site: challenges and benefits. J Usabil Stud 7(1):9–30

Turner C, Lewis J, Nielsen J (2006) Determining usability test sample size. In: Karwowski W (ed) International encyclopedia of ergonomics and human factors. CRC Press, Boca Raton

Wentz B, Cirba M, Kharal N, Moran J, Slate M (2012) Evaluating the accessibility and usability of blogging platforms for blind users. In: Langdon PM, Clarkson PJ, Lazar J, Heylighen A (eds) Designing inclusive systems. Springer, London

Designing Accessible Workplaces
for Visually Impaired People

J. L. Gomez, P. M. Langdon, J. A. Bichard and P. J. Clarkson

1 Introduction

In the new digital era, workplaces are constantly changing to incorporate, e.g. new information and communication technologies (ICTs), as well as ergonomic features that attempt to improve the wellbeing of workers. Such technological advances, along with globalisation and demographic changes (i.e. ageing populations, falling birth rates and migration) have modified the world of work that we used to know: organisations have increasingly capitalised on ideas, creativity and potential contributions of their employees (Burke and Ng 2006). Despite this, the Employment Forum on Disability (EFD 2008) highlighted that some changes (e.g. computers, work stations and training), which should make working conditions easier for employees, often create greater barriers for workers with a disability. In this context, Greisler and Stupak (2002) indicate that, for example, many industrialised countries see technological progress 'as a ready means through which governments can address issues of social exclusion'. However, inaccessible technologies may be a cause of major exclusion in the workplace for disabled workers who cannot interact with such technologies (Foster 2011) and with the built environment.

Scholars do agree that the current legal instruments and economic development models adopted within the global community have been insufficient to improve the welfare and inclusion of disabled people (Braithwaite et al. 2008). This is likely to be because the majority of countries approach disability as segregated division rather than embracing human diversity. Indeed, human beings are divided into

J. L. Gomez (✉) · P. M. Langdon · J. A. Bichard · P. J. Clarkson
Engineering Design Centre, Department of Engineering, University of Cambridge,
Cambridge, UK
e-mail: jlg55@eng.cam.ac.uk

P. M. Langdon
e-mail: pml24@eng.cam.ac.uk

P. M. Langdon et al. (eds.), *Inclusive Designing*, DOI: 10.1007/978-3-319-05095-9_24,   269
© Springer International Publishing Switzerland 2014

those who belong to the norm and 'others' whose degree of otherness follows from the severity of their impairment (Clapton and Fitzgerald 1997). As a result, disabled peoples' needs, so far, have been mostly covered by charities, welfare states and other supports, like sheltered workshops, which are only useful in keeping disabled people out of the social mainstream, instead of ensuring a real right to employment (Oliver and Barnes 1998).

On the other hand certain disabled movements, such as wheelchair users, have indeniably exerted a greater influence to avoid design exclusion. Organisations are currently taking further action to increase awareness of the design barriers faced by the visually impaired in workplace. This paper is intended to explore the field of Inclusive Design (ID) in the UK, design for all (DfA) in Europe and Asia and universal design (UD) in the US. How can we reduce design barriers faced by low vision and blind people in the workplace? The aim is to identify disabling barriers as well as provide specific design standards to improve workplace accessibility for the visually impaired.

The first part of this paper will outline how the theoretical construction of disability was linked with insights in the historical, economic and political evolution of industry and activist movements leading the design mainstream. An overview of this evolution will contribute to a better understanding of the myths and realities embodied in non-disabled people's view of disabled people's capabilities. With this in mind, the second part of this paper will give some guidance on supporting visually impaired people's needs and how DfA may help to do this in the workplace.

2 Industrial Evolution and the Theoretical Construction of Disability

2.1 Models of Disability

The worldwide tendency is to stigmatise people with disabilities by their reduced capabilities. In this way, able-bodied people have stereotyped disabled people (Scullion 2009) by creating labels that not only classify them as incapable, unproductive and inefficient within the virtual, social and physical workplace, but also reduce them to a lower status at work (Finkelstein 1980; Oliver 1980; Barnes 2011). Disabled people are victims of what can be called 'functional stigmatisation', i.e. they are considered out of 'the norm' or 'the average population' in not possessing 'the abilities that lie within normal ranges of behaviours for a person at a particular chronological age' (Bichard et al. 2007). Such stigmatisation as 'abnormal' results from an historical conceptualisation of disability, strongly influenced by factors such as religion, industrialisation, free market and liberal philosophy, as well as medical and scientific progress (Finkelstein 1980; Oliver 1980).

As a result, scholars highlight that disability has been characterised worldwide by several distinct models throughout different periods of history.

2.2 Disability, Industrialisation and Design

In the mediaeval age (feudalism), disabled people, despite religious superstitions, were able to work in rural areas and/or in small-scale industry. With the arrival of industrialised production and the worldwide rise of capitalist societies, the organisation of work changed and people were forced to fit into new types of paid work in order to survive, moving from rural cooperative systems towards urban factories (Russell and Malhotra 2002).

The growing complexity of such social changes along with the development of a new industrial class excluded those who possessed any impairment, and so the idea of disability was born (Russell and Malhotra 2002). In general, before the concept of disability, in which impairment was considered a specific 'abnormality of the body', religious people used to believe that people with impairments were the result of sin or demons. The disabled were left to fend for themselves or hidden by their families (Miles 2002). With the introduction of scientific innovation, dogmatic belief in religious authority started to diminish, and so the new medical approach was adopted to find solutions to disability issues. The rapid development of medicine and advancements in medical diagnosis, procedures and technology led to an increasing sense of confidence in medical science. So, disability was conceptualised around people's inabilities due to impairment, which transformed the individual and social image of impairment. In the 1950s, after two world wars, disability activists started to create an awareness that disability was out of the body, not to be understood as an individual problem lying in the inability of the body (Sieber 2008). They developed the concept that 'disability was within the built environment' (Oliver 1990). Despite this new view of disability, most European industrialised countries expanded the welfare state to compensate for poor opportunities given to disabled people, increasing the provision of social services and regulating their lives of the recipients. Following this evolution of liberal markets and social organisations, mainstream design contributed to increase the exclusion of disabled people who, to this date, still live in a world where they are mostly neglected by their non-disabled peers. Nonetheless, design thinking has moved from standardised production towards a growing awareness of human-centred design: see Fig. 1.

End of 19th century

- The body became significant to participate in society and be economically active (Oliver, 1980).
- People needed to function within the range considered normal to achieve tasks with minor flexibility, major speed, bigger machines and minimal wasted time (Finkelstein et al., 1993).
- Impairment started to be medically classified into categories such as: motor, mental, sensory, and cognitive (Shullion, 2009).
- Impairments "could be overcome through the adaptation of the body itself" (Herssens, 2013).
- Design for the average user.
- Design thinking was mainly focused on supporting fast standardised manufacturing processes and covering the needs of average users (able-bodied people) (Finkelstein, 1980).

50s

- The idea was that the impaired body must be restored, adapted and cured (Scullion, 2009).
- A growing number of people were injured (impaired) by the introduction of new machineries and two world wars.
- A new social discourse began a process of change in public policies and design practices (IHCD, 2008).
- Rehabilitation engineering and assistive technology emerged to improve prosthetics and artificial limbs as well as special products to improve communication, mobility and transportation.
- Important social changes started to outline a new approach to disability legislation and social research.

60s – 70s

- Impairments slowly started to be understood as a problem of the whole society (Oliver, 1980).
- Disabilities were supported by assistive technologies such as personal devices to cover physical, sensory, and cognitive needs (Barnes, 2011).
- Design thinking was to help disabled people to function as independently as they could.
- One of the most common design approaches used was Free Barrier Design or Specific Needs Design (Heiss et al., 2010).

80s

- People with impairments could not be forced to adapt and fit into the world, but rather the designed world must be created to fit to its users (Imrie, 2004).
- Disability was in barriers socially imposed by able-bodied people (Oliver, 1980).
- Product manufacturers recognised the market-broadening potential of more accommodating products (IHCD, 2008).
- Design focus was Free Design Barriers or Specific Needs Designs which was strongly focussed on wheelchair users.
- Design thinking was to reduce barriers and meet disabled people's specific needs.
- The US introduced accessible design, which had a similar purpose to Free Barrier Design, but has been seen as a more inclusive way to design. However, the requirements made are only supporting legislation (Wolfgang and Ostroff, 2001).

90s

- Globalisation of the marketplace changed forms of business strategy competition and consumer diversity became a key factor for success (EIDD, 2008).
- Design thinking needed to be more sensitive to individual differences without restrictions of ability, age, sex, ethnicity, etc.
- Design approach aimed at finding solutions for as many people as possible (Barnes, 2011), and with "a major perceptual orientation to humanity" (Davies and Lifchez, 1987).
- Disability is in the environment, out of the impaired body (Sieber, 2008).

21st century

- The current generation of children, baby boomers entering middle age, older adults, people with disabilities, and individuals inconvenienced by circumstance, constitute a market majority (IHCD, 2008).
- Design has started to value, recognise and respect all users (Barnes, 2011).
- Emergence of a new design thinking known as inclusive design in the UK, universal access in the US, and design for all.
- Design thinking moves to cover new user demands to enable people to live in a world with no disabling barriers that impact on their daily lives (Herwig, 2008).
- Products, facilities, devices, services and programs must be designed to serve an increasingly diverse clientele (IHCD, 2008).

Fig. 1 Evolution of attitudes to design and disability

3 Data on Visually Impaired People

Statistics are usually considered important for planning purposes and for advo-
cating peoples' rights. A review of the literature confirmed that an individual with
a visual impairment is someone who has sight loss that cannot be corrected by
glasses or contact lenses, and which restricts their ability to successfully complete
day-to-day activities. The impact of sight loss on the way in which visually
impaired people function depends on whether their disability is congenital or
adventitious. A congenital disability is one which occurs at birth or immediately
after, so a visual memory has not developed. Adventitious refers to a hereditary
condition or trauma that occurs after normal vision and visual memory develop
and can remain in the brain.

In the UK, for example, someone is legally blind if he/she cannot do any work
for which eyesight is essential (NHS 2013) after their visual acuity and visual field
have been measured. Worldwide, the common and more accepted definition of
blindness and low vision is given in the International Classification of Diseases,
ICD (Updated and reviewed in 2006). The ICD established four levels of visual
function: normal vision, moderate visual impairment, severe visual impairment
and blindness. Moderate and severe impairment are grouped in the term 'low
vision' which together with blindness constitutes the whole spectrum of visual
impairment (WHO 2012). In the UK visually impaired people are categorised into
three main groups (NHS 2013):

1. People who have extremely poor visual acuity (less than 3/60) but have a full
   field of vision.
2. People who have poor visual acuity (between 3/60 and 6/60) and a severe
   reduction in their field of vision.
3. People who have average visual acuity (6/60 or better) and an extremely
   reduced field of vision.

The World Health Organization (WHO 2012) states that blind and partially
sighted people make up to 285 million around the world. Of these, a greater
number are low vision (246 million), with only 39 million blind people. The WHO
also reports that 90 % of visually impaired people live in developing countries and
that 80 % of impairments can be avoided or cured. Most reports on visually
impaired people point out that the vast majority are unemployed with few
opportunities to change their condition due to, e.g. lack of access to education,
assistive technologies and accessible environments. Only 2 % of visually impaired
children in low-income countries have access to higher education and just 5 % of
this 2 % will have access to employment. In rural areas, this situation is even
worse since up to 100 % of children are out of the educational system and
employment. In industrialised countries the situation is not very much better. Most
blind people in these countries have reported that their lives are affected more by

lack of opportunities than lack of ability. They do not have the opportunity to go to school, obtain work or enjoy family and social life. Unemployment rates go from 75 % (which is the average), to 50 % at best.

4 Applying Inclusive Design to Eliminate Barriers Within the Workplace

4.1 Relationship Between Inclusive Design and Barriers Faced by Visually Impaired People

Everything around us implies a design, from the shape of our tables to the access of computer functions through software. Margolin (2008) states that design is the thinking, idea or plan that organises all levels of production. To date, disability has been largely neglected in the design of the world in which we live. Unfortunately, the exclusion of disabled people has also resulted in their able-bodied counterparts not considering them in their design processes (Roosevelt 2006). Most products and services have not been designed to facilitate the interaction of all human beings; in fact they strongly exclude those who have a physical, mental or sensorial impairment. In simple words, most of the material world has been wrongly designed, and does not enable everybody to interact properly with objects, artefacts or environments.

ID is a framework that aims to design for the widest range of people operating in the broadest range of situations without special or separate adaptations (Clarkson 2003). It is human-centred design of everything with everyone in mind. It is, so far, the best known principle to lead the development of products and services in a direction that optimises their usability and accessibility. However, it is a concept unknown to a great part of society, with little evidence of its use to the workplace.

ID has its roots in the principle that designing for the minority also works for the majority (Imrie and Hall 2004). Thus it can completely eliminate disabling barriers in the current built world (Barnes 2011).

Some scholars, against this design mainstream, think that the need for individualised accommodations cannot be eliminated in some situations. However, they agree that ID provides a set of key principles and guidelines needed to develop and implement effective accommodations that will work for all employers and employees (Loy and Carter 2007). Despite these debates most scholars agree that, in the workplace, an analysis of the needs, wishes and requirements of the workers (Timmon et al. 2009) would help to create a more accessible, safe and friendly space in three broad areas (DBTAC Northwest 1997):

1. The working environment, including the employees' workstations as well as the entire work facility or worksite.

2. Technologies and tools, e.g. computers and communication technologies, manufacturing tools, controls and equipment, furniture and safety equipment.
3. Work policies, interaction, communication, safety and most importantly for the accommodation process, the methods used to complete work tasks.

4.2 Design and the Visually Impaired Worker

Visually impaired people are the subject of extensive research, but this has been criticised for providing guidance and instructions which are not aligned with the real users' needs and language. These common mistakes result in inadequate solutions, which do not take peoples' visual *functioning* into consideration. In simple words, visually impaired people who *medically* have been diagnosed with the same visual condition (the same visual acuity or visual field) are likely to use their sight quite differently (Nicolau et al. 2009). This is because a person's visual functioning is affected not only by a decrement of their visual acuity and visual field but also by the lack of reference points and visual cues. For example, someone who has congenital blindness will have different visual functioning from someone who has adventitious blindness.

Therefore, if visually impaired people are the design focus, the first step to effectively design an accessible workplace would be to get feedback from visually impaired users themselves rather than professional people who work with them, as is usually the case. Visually impaired people can, in this way, help designers to understand vision difficulties such as sensitivity to light or glare, blind spots in their visual fields, fatigue, variations in visual function as a result of changing emotions in unknown environments, information missed for lack of audible or tactile signals, etc. In other words, designers are required to analyse how the visually impaired create their own mental maps, verbalise an image and communicate with others. They must comprehend how they explore environments, products, artefacts and places, as well as their difficulties, capabilities, needs and behaviours.

In general, visually impaired peoples' needs at work are not well documented, and this makes it harder to ensure accessibility in the workplace. Designers need to know how visually impaired workers use their vision and understand that there is no common rule for all those who are low vision or blind. Most of the time, the knowledge acquired by designers is not accurate since it comes from specialists who are not disabled or from certain literature which has a deeper focus on physical disability, rather than the specific needs of those who suffer loss of sight. Although progress has been made in the last few years, many barriers faced by the visually impaired are unknown to sighted people, but with more knowledge, they would easily be broken down.

These barriers can be overcome if we keep in mind that the other senses do not fully compensate for the loss of sight. Sight is valued as the most important sense

Outdoor and indoor spaces	Workstation and safety	Communication and information
• The design of spaces affects how the visually impaired travel and move from one place to other. Thus simple layouts of the travel path, with for example, no obstacles and entrances similarly designed will be easily recognised and remembered. • Safety standards must be ensured through, for example, multi-sensory alarm signals (auditory, visual); emergency and safety equipment clearly identified and placed in a conspicous location. • Door handles, light switches, lift controls, taps, etc. should be standardised in function and position. • Low vision people can see contrasts in colour and light which are important to facilitate references (for example paintwork with different colours) and negotiate unknown places.	• Worksite accommodation will be quite varied. However there are some basic best practices that can lead to the most effective adaptations. These include minimal glare, blinds or curtains on windows adjacent to workstations (ergonomic furniture), storage in areas that are reachable for all employees (not high up or over a worker's head). The key is in designing so that the visually impaired workers can act independently. Thus a helpful design will, for example, use sliding doors so they do not project if left open. The workstation must be free of obstacles such as badly positioned furniture.	• The most important thing is to ensure that visually impaired workers are given access to all information given to other workers. Therefore print text, for example, has to be adapted for them. This means that non-disabled workers need to know what adaptive device, method or equipment (assistive technology) the visually impaired worker is using. Here are some examples: • Portable scanner to transform text into audio or a compatible format to be read on a computer with a screen reader. • Screen reader to transcribe text on a computer screen to audible speech. • Screen magnifier software to increase the size of text and images on a computer screen. • Portable magnifier to increase the size of text and images on printed documents. • Closed circuit television to enlarge an image on a computer and project it onto a screen. • Telescopes to view distant objects.

Fig. 2 Basic design standards for visually impaired workers

to learn and explore our environment. It is through sight that much information is received and processed. Touch and hearing can be ineffective substitutes, although people with visual impairments learn compensatory skills and adaptive techniques that help them to deal with their loss of sight (GNL 2001):

• Low vision—individuals still have poor vision as their primary sensory channel to explore the environment.
• Functionally totally blind—individuals have limited vision for functional tasks, but they use their tactile and auditory channels for learning.

Designers must never forget that in the workplace, visually impaired workers have more difficulties sharing common visual experiences with their sighted peers. They need an environment that they can easily remember (See Fig. 2). The restricted ability to see around them reduces their motivation to explore the environment, initiate social interaction and manipulate unknown objects. Blind and low vision people therefore miss important information which slows their familiarisation with daily tasks at work and they are not able to integrate new learning as quickly as expected. For this reason, visually impaired workers need specialised instruction in a number of essential areas such as how to use corporate software with a screen reader, mobility and orientation in the workplace, changes in furniture position, etc. Without this, they cannot be independent at work. In this context, it is relevant that the visually impaired learn at roughly the same rate as others, but require direct interventions orient them among the objects in their environment (Jacko and Sears 1998).

5 Conclusion

For many years, negative attitudes and discriminatory employment practices against disabled people arose from predominant models used in portraying disability. Finkelstein (1980) states that such models have produced beliefs about disability that emerged from what he defines as 'historical materialism'. As a result, industrial, economic and scientific development has provoked a profound isolation and exclusion of disabled people from mainstream life and specifically from the workforce.

Disability movements around the world claim that states and society have taken an ableist, patriarchal and protectionist perspective, which put them at a disadvantage in competing with the able-bodied. Indeed, most evidence demonstrates that able-bodied people have not yet recognised that environments are the main course of disability. And this will not be possible while they continue to focus on what isolates disabled people rather than what is common to all human beings (Berman 2005).

In light of the above, the World Bank states that countries must replace existing welfare states by a more inclusive approach (Braithwaite et al. 2008), that values the diversity existing in the modern world (Zoellick 2007) to cover disabled peoples' needs. However, achieving inclusivity for all citizens means not only recognising the values and abilities of each person, but also including them as an important part of decision making in order to facilitate this more inclusive development (Snider and Takeda 2008). In this context, when visually impaired people are the design focus, the principles of DfA/ID are being indicated as a solution to reduce current disabling barriers. However, the key to its success will be for organisations to incorporate it adequately into their regular design practices (DBTAC Northwest 1997; Carter 2007; EIDD 2008). Beyond rhetorical concerns over ensuring universal access and right to work for disabled people, if the goal is to create a more inclusive workforce, the first step unquestionably is to create a standard, that covers the needs of the visually impaired at work.

Echoing this, a wide range of products can be designed to meet visually impaired people's needs, but many designers argue that it is quite hard to incorporate ID in their work and sometimes it is impossible to achieve full integration for all human beings. However, although there will always be smaller groups of people excluded by inaccessible design, still it is possible to expand research to enable the integration of as many visually impaired people as possible.

References

Barnes C (2011) Understanding disability and importance of design for all. J Accessibility Des All 1(1):54–79

Berman A (2005) Desarrollo inclusivo: un aporte universal desde la discapacidad. Banco Mundial, Latinoamericana y El Caribe

Bichard J, Langdon PM, Coleman R (2007) Does my stigma look big in this? Considering the acceptability and desirability in the inclusive design of technology products. In: Proceedings of the 4th international conference on universal access in human-computer interaction. Beijing, P.R. China

Braithwaite J, Carroll R, Peffley K (2008) Disability and development in the World Bank: FY2007-2007. World Bank, Washington

Burke R, Ng E (2006) The changing nature of work and organisations: Implications for human. Resour Manag Rev 16(2006):86–94

Clapton J, Fitzgerald J (1997) The history of disability: a history of otherness. New Renaissance Magazine 7(1)

Clarkson PJ (2003) Inclusive design for the whole population. Springer, London

Davies C, Lifchez R (1987) An open letter to architects. In: Lifchez R (ed) Rethinking architecture: design students and physically disabled people. University of California Ltd, Los Angeles

DBTAC Northwest (1997) Universal design in the workplace. DBTAC Northwest Disability and Business Technical Assistance Center Americans with Disability Act (ADA) Information Center. http://www.dbtacnorthwest.org/_public/site/files/ada/documents/UD_Workplace_Final_02.pdf. Accessed 31 Oct 2013

EIDD (2008) Design for all Stockholm declaration 2004. European Institute on design and disability. http://www.designforalleurope.org/Design-for-All/. Accessed 31 Oct 2013

Finkelstein V (1980) Attitudes and disabled people: issues for discussion. World Rehabilitation Fund, New York

Herwig O (2008) Universal design: solutions for barriers-free living. Birkhäuser Verlag AG, Basel

Herssens J (2013) Designing for more– towards a global design approach and local methods. In: Proceedings of the include Asia 2013, Hong Kong Design Centre, Hong Kong, P.R. China

IHCD (2008) History of design for all. Institute for Human Centered Design. www.adaptiveenvironments.org. Accessed 5 June 2010

Imrie R, Hall P (2004) Inclusive design: designing and developing accessible environments. Taylor and Francis Group, US

Jacko J, Sears A (1998) Designing interface for an overlooked users group: considering the visual profiles of partially sighted users. In: Proceedings of the 3rd international ACM conference on assistive technologies, Marina del Rey, CA, US

Loy B, Carter L (2007) Improving the workplace one accommodation at a time. Job Accommodation Network, US

Margolin V (2008) Design discourse: History, theory, criticism. University of Chicago Press, US

Miles M (2002) Some influences of religion on attitudes toward disabilities and people with disabilities. J Relig Disabil Health 6(2/3):117–129

NHS (2013) Visual impairment. http://www.nhs.uk/Conditions/Visual-impairment/Pages/Introduction.aspx. Accessed 31 Oct 2013

Nicolau H, Guerreiro T, Joaquim J (2009) Designing guides for blind people. Departamento de Engenharia Informatica, Instituto Superior Tecnico, Lisboa

Oliver M (1980) The politics of disablement: a sociological approach. MacMillan, London

Oliver M (1990) The Politics of disablement. Macmillan, Basingstoke

Oliver M, Barnes C (1998) Social policy and disabled people: from exclusion to inclusion. Longman, London

Roosevelt T (2006) Building on the promise of diversity: how we can move to the next level in our workplaces, our communities, and our society. AMACON, US

Russell M, Malhotra R (2002) Capitalism and disability. Socialist Reg 38:211–228

Scullion P (2009) An analysis of the 'six disabled lives' in the UK: important or impotent? Int J Ther Rehabil 16(7):385–390

Siebers T (2008) Disability theory (corporealities: discourses of disability). University of Michigan Press, Michigan

Snider H, Takeda N (2008) Design for all: implications for bank operations. World Bank, Washington

Timmon J, Fesko S, Hall A (2009) From diversity to inclusion: considering the universally designed workplace. OEEO Diversity Bizz. http://www.diversityinc.com. Accessed 8 Feb 2011

WHO (2012) Visual impairment and blindness. World Heath Organisation. http://www.who.int/mediacentre/factsheets/fs282/en/. Accessed 31 Oct 2013

Zoellick R (2007) An inclusive and sustainable globalisation. World Bank, Washington

Author Index

P. M. Langdon et al. (eds.), *Inclusive Designing*, DOI: 10.1007/978-3-319-05095-9,
© Springer International Publishing Switzerland 2014

CPSIA information can be obtained
at www.ICGtesting.com
Printed in the USA
LVHW02*2255101217
559299LV00001B/92/P

9 783319 050942